S. HRG. 115–67

EXTREME WEATHER AND COASTAL FLOODING: WHAT IS HAPPENING NOW, WHAT IS THE FUTURE RISK, AND WHAT CAN WE DO ABOUT IT?

FIELD HEARING

BEFORE THE

COMMITTEE ON COMMERCE, SCIENCE, AND TRANSPORTATION UNITED STATES SENATE

ONE HUNDRED FIFTEENTH CONGRESS

FIRST SESSION

APRIL 10, 2017

Printed for the use of the Committee on Commerce, Science, and Transportation

U.S. GOVERNMENT PUBLISHING OFFICE

26–599 PDF   WASHINGTON : 2017

For sale by the Superintendent of Documents, U.S. Government Publishing Office
Internet: bookstore.gpo.gov   Phone: toll free (866) 512–1800; DC area (202) 512–1800
Fax: (202) 512–2104   Mail: Stop IDCC, Washington, DC 20402–0001

SENATE COMMITTEE ON COMMERCE, SCIENCE, AND TRANSPORTATION

ONE HUNDRED FIFTEENTH CONGRESS

FIRST SESSION

JOHN THUNE, South Dakota, *Chairman*

ROGER F. WICKER, Mississippi	BILL NELSON, Florida, *Ranking*
ROY BLUNT, Missouri	MARIA CANTWELL, Washington
TED CRUZ, Texas	AMY KLOBUCHAR, Minnesota
DEB FISCHER, Nebraska	RICHARD BLUMENTHAL, Connecticut
JERRY MORAN, Kansas	BRIAN SCHATZ, Hawaii
DAN SULLIVAN, Alaska	EDWARD MARKEY, Massachusetts
DEAN HELLER, Nevada	CORY BOOKER, New Jersey
JAMES INHOFE, Oklahoma	TOM UDALL, New Mexico
MIKE LEE, Utah	GARY PETERS, Michigan
RON JOHNSON, Wisconsin	TAMMY BALDWIN, Wisconsin
SHELLEY MOORE CAPITO, West Virginia	TAMMY DUCKWORTH, Illinois
CORY GARDNER, Colorado	MAGGIE HASSAN, New Hampshire
TODD YOUNG, Indiana	CATHERINE CORTEZ MASTO, Nevada

NICK ROSSI, *Staff Director*
ADRIAN ARNAKIS, *Deputy Staff Director*
JASON VAN BEEK, *General Counsel*
KIM LIPSKY, *Democratic Staff Director*
CHRIS DAY, *Democratic Deputy Staff Director*
RENAE BLACK, *Senior Counsel*

CONTENTS

	Page
Hearing held on April 10, 2017	1
Statement of Senator Nelson	1

WITNESSES

Hon. Ted Deutch, U.S. Representative from Florida	4
Hon. Jeri Muoio, Mayor, West Palm Beach, Florida	5
Hon. Paulette Burdick, Mayor, Palm Beach County, Florida	7
Ben Kirtman, Ph.D., Director, Cooperative Institute for Marine and Atmospheric Sciences, and Director, Center for Computational Science, Climate, and Environmental Hazards, University of Miami	9
Prepared statement	10
Leonard "Len" Berry, Ph.D., Emeritus Professor, Geosciences, Florida Atlantic University, and Vice President, Government Programs, Coastal Risk Consulting, LLC	19
Prepared statement	20
Carl G. Hedde, CPCU, Head, Risk Accumulation Department, Munich Reinsurance America Inc.	35
Prepared statement	36
Dr. Jennifer L. Jurado, Chief Resilience Officer and Director, Environmental Planning and Community Resilience Division, Broward County, Florida	39
Prepared statement	41

APPENDIX

Response to written questions submitted by Hon. Bill Nelson to:

Ben Kirtman, Ph.D.	55
Leonard "Len" Berry, Ph.D.	55
Carl G. Hedde	56
Dr. Jennifer Jurado	57

EXTREME WEATHER AND COASTAL FLOODING: WHAT IS HAPPENING NOW, WHAT IS THE FUTURE RISK, AND WHAT CAN WE DO ABOUT IT?

MONDAY, APRIL 10, 2017

U.S. SENATE,
COMMITTEE ON COMMERCE, SCIENCE, AND TRANSPORTATION,
West Palm Beach, FL.

The Committee met, pursuant to notice, at 1:38 p.m., City of West Palm Beach Commission Chambers, 401 Clematis Street, West Palm Beach, Florida, Hon. Bill Nelson presiding.

Present: Senator Nelson [presiding].

OPENING STATEMENT OF HON. BILL NELSON, U.S. SENATOR FROM FLORIDA

Senator NELSON. The meeting of the Senate Committee on Commerce, Science, and Transportation will come to order.

This is a field hearing that is being held by the Senate Committee. We have had these kinds of field hearings many times before, but on the subject of the day, extreme weather and coastal flooding—in other words, sea-level rise and climate change—we had an identical hearing 3 years ago in what was then ground zero in Miami Beach. And so this record will be compiled with the other records of the Committee that are occurring in Washington on this topic that we will be hearing our witnesses discuss today.

I want to welcome all of you.

If we may first have Reverend Gerald Kisner of Tabernacle Missionary Baptist Church, if he will give us our Invocation?

Reverend KISNER. Thank you, Senator and Mayor. Let us pray.

God of Creation, Creator, again, we come into your divine and magnificence presence. First of all, thank you for the lifegiving force that you give to each and every one of us. We thank you that you are a God of creation. We pray, Lord, even as these deliberations and these testimonies are lifted up, that all will be done to help maintain our job as the custodians of your creation, be with all, that we might do the thing that is pleasing in thine sight. And, Lord, as always, we will be careful to thank you. We lift up this prayer, the only one that matters. Amen.

Senator NELSON. Thank you, Reverend Kisner.

Now, we have a treat. For our Pledge of Allegiance, we have students from The Greene School, and they are going to come up and lead us in the Pledge of Allegiance.

So, students, if you can come up?

[Pledge of Allegiance.]

Senator NELSON. Thank you so much.

Thank you to everybody for coming. This is an extraordinary turnout. There is obviously a lot of interest in the subject matter, and we appreciate so much your being here. I want to welcome you, as I said, to an official meeting of the Commerce, Science, and Transportation Committee. We are doing this because we sit at ground zero of the impacts of climate change in the U.S.

While there are still some who continue to deny that climate change is real, South Florida offers proof that it is real and it is an issue we are going to have to confront in the decades ahead. All of us here today know that Florida is particularly vulnerable to the effects of climate change. We have over 1,200 miles of coastline, more than any other state in the continental U.S., and over three-quarters of our state's residents live in coastal areas.

Florida is also quite flat. The highest point in the peninsula of Florida is in the center of the state near Lake Wales at Bok Tower, and it is only 345 feet high.

Our communities are already experiencing regular, nuisance flooding, especially during the king tides.

You can see a photograph taken in Miami Beach in 2015. Here is a gentleman trying to walk across the street. That is just a year and a half ago. Obviously, the floodwater there reaches above the curb. And it has gotten so extreme that, if you look at this photograph, you will see the sea creatures are showing up in bizarre places. Here is an octopus in a parking garage.

The National Academy of Sciences found that 67 percent of the nuisance floods in the U.S. are being driven by human-caused global sea-level rise. In Miami Beach, tide-induced flooding has increased by more than 400 percent in the last decade. In southeast Florida, sea-level rise has tripled since 2006.

The resulting impacts of coastal flooding, saltwater intrusion, storm surge, and land erosion on the Florida coast has prompted local governments to act. And that is one of the reasons we wanted to bring a field hearing of the Commerce Committee here to South Florida, because you will be hearing from some of those local governments.

Here in Palm Beach County, more than 20 acres of beach and sand dunes had to be restored following Hurricane Sandy to better protect shoreside communities from flooding and severe weather.

Observations such as these—not models, not projections, but data—tell us that the average global sea-level rise is happening.

We had a NASA scientist 3 years ago in Miami Beach. You may have read about this scientist, because he was the scientist who was also an astronaut, and he worked until his last moment that he could function, since he had incurable pancreatic cancer. He testified 3 years ago that measurements—that is not forecast, that is not projections—measurements over the last 40 years were showing 5 inches to 8 inches in South Florida of sea-level rise.

So the rising ocean temperatures also have been linked to increasing hurricane activity and intensity as hurricanes draw more energy from the warm water.

Now you wonder, why is all of this happening? As we emit more and more greenhouse gas, specifically carbon dioxide, CO_2, and, in some cases, sulfur dioxide, SO_2, and those gases rise, they create in the upper atmosphere the greenhouse effect.

So if you are familiar with a greenhouse, it has a ceiling of glass. It lets in the sunlight but then it traps the heat, thus a greenhouse. So too, without the pollutants in the upper atmosphere, the sun's rays would come and reflect off of the Earth. Some of the heat would be retained but a lot of them radiate back out into space. When you put a ceiling up there, a ceiling of CO_2 and SO_2, it acts like a greenhouse glass top. It traps the heat from radiating out into space and, thus, starts to heat the Earth.

Two-thirds of the Earth is covered with water. What happens to water when it heats up? It expands. And, thus, you see the phenomenon that is happening in this country.

In 2016 alone, there were 15 weather and climate disaster events that caused over $1 billion in damage each and resulted in 138 deaths across the country. In the first three months of this year, there have been five more events with losses exceeding $1 billion. These included a flood, a freeze, and three severe storms. Five is the largest number of billion-dollar events for January to March ever recorded.

Sea-level rise will worsen the issue by creating deeper waters near shore, causing higher waves and stronger storm surges, especially during hurricanes. This is especially concerning considering that 79 percent of Florida's economy is generated in coastal communities, and over $130 billion in beach real estate is at risk.

So what can we do about it? Well, first, we need to be clear about the facts that are presented to the public and fight against the political censorship of our climate scientists and their data. If a doctor were barred from using the word "cancer," he or she cannot do his job. The same is true with scientists and the work that they do to understand and educate the public about the Earth's increasing fever.

Now if you think I am kidding, I want you to know that I have gone out to some of the agencies of the Federal Government and they are being warned not to use the term "climate change." And of course, we have had our own experience here in the state of Florida by a similar message being given by the Administration in this state to State employees of not using the term "climate change."

So at the end of the day, what we have to do is limit the greenhouse gas emissions, but we must also create more resilient communities. And I want to take the opportunity to applaud, to really congratulate and to thank and to applaud Palm Beach, Broward, Miami-Dade, and Monroe counties for their work on the Southeast Florida Regional Climate Change Compact and the City of Punta Gorda on the West Coast and the City of Satellite Beach further north on the East Coast for their sea-level rise adaptation planning and their efforts to become more resilient communities.

And I know that they will keep up this work, and I hope that others will follow their lead. And I believe that, at the Federal level, we should be providing more tools to these communities, not less.

Now, lest you think that we are all in a period of partisan warfare and ideological extremism, which someone could conclude that by watching the news, I want you to know that, in the past two weeks, I have had two separate conversations with Vice President Pence on trying to get a bipartisan infrastructure bill going. Of course, the infrastructure that you will hear about in Miami Beach and the South Florida compact is extremely expensive infrastructure. Miami Beach has already spent millions and millions of dollars on the big, big pumps to get that water out of the streets when it comes, and to raise the level of the streets.

So, before I introduce the distinguished panel, I want to turn to our colleagues here, and I want to hear from them. Before I do, let me introduce the elected officials that are here in the audience. Hold your applause, please.

If you all will stand and remain standing, we can recognize you all at once.

County Commissioner Melissa McKinlay of Palm Beach County; Martin County Commissioner Doug Smith; Martin County Commissioner Ed Fielding; Mayor of the Town of Palm Beach, Gail Coniglio—thank you, Mayor; Riviera Beach Councilwoman Dawn Pardo; Kim Ciklin, representing County Commissioner Hal Valeche; Marian Dozier, representing State Senator Bobby Powell; Charity Lewis, representing Congresswoman Lois Frankel; Jervonte Edmonds, representing Florida House Representative Al Jacquet; West Palm Beach Commissioner Materio; Congressman Hastings being represented by Dan Liftman; County Commissioner Dave Kerner, I think Danna White is representing him; and Lori Berman, our State Representative.

Now let's welcome them.

[Applause.]

Senator NELSON. All right, we have a star-studded panel up here. I want you to know that I, as is appropriate and respectful, even though I am the only Senator here from the Commerce Committee, I invited the entire Congressional delegation from South Florida, Republicans and Democrats. And for one reason or another, and I make no inferences here—this is the Easter recess and Passover starts tonight at sundown. So there were a number of folks that just simply could not be here.

But I am very pleased that one of the South Florida delegation, Congressman Ted Deutch, is here.

So, Congressman, your opening comments?

STATEMENT OF HON. TED DEUTCH, U.S. REPRESENTATIVE FROM FLORIDA

Mr. DEUTCH. Thanks very much, Senator Nelson. I will be heading out as soon as this is over to help prepare matzo ball soup.

[Laughter.]

Mr. DEUTCH. I really appreciate your holding this hearing here in Palm Beach County. Climate change and coastal flooding is not a threat to our future in South Florida. It is, as you know, a threat today.

The challenges require immediate action. I want to thank you for bringing us together for this critical hearing.

I also want to thank our witnesses, the private sector, the public sector, universities all well represented here and speak to the kind of broad-based approach that we are taking here in South Florida and nationally.

Recently, NOAA confirmed that 2016 was the warmest year on record, something that we have heard but need to do something about.

Rising sea levels and coastal flooding are forcing real estate lenders to consider sea-level rise before issuing 30-year mortgages. Recent studies have predicted that one million Florida homes will be underwater by the year 2100 if we do not act. That translates to one in 8 homes, which would take a devastating toll on the real estate market, $413 billion in losses, if we do not take this seriously.

In South Florida, we have sea-level rises 10 times the global rate. Our porous limestone peninsula fills with water that floods homes and spills over seawalls. The king tides, as you pointed out, Senator, are bringing floodwaters into our streets on sunny days each year.

And that is why this area is so important. We have to be prepared with mitigation plans and investments in adaptation.

And our local leaders cannot do it alone. Reducing the impact of climate change is going to require interaction at a serious level between Federal, State, local, and the private sectors to rise above those challenges to find common ground.

That is why, last year, in the House, we founded a bipartisan climate solution caucus, which is a first of its kind in Congress. We are focused on building our bipartisan membership. We have a good showing already from Florida Congressman Ros-Lehtinen; Congressman Mast; Congressman Crist; Carlos Curbelo, the Congressman from Miami-Dade, who is my Co-Chair; and there are lots of other Members who are anxiously waiting to get on, but can only join when a Democrat and a Republican join together.

In total, we have pulled 38 Members thus far aboard evenly split. Seventeen states are represented from diverse regions, each reflecting the varied challenges presented by climate change.

This hearing today gives us the opportunity to explore these issues in a very public way to make a record, Senator Nelson, that we are grateful that you will make and take back to Washington to share with your fellow members of the Commerce Committee so that, hopefully, we can start to see the sort of coming together that is necessary, and that is starting today here, thanks to you.

I appreciate it very much. Thanks, Senator Nelson.

Senator NELSON. Thank you, Congressman.

We are here at the kind invitation of the City of Palm Beach, and we are so pleased that you could join us, Mayor Jeri Muoio.

STATEMENT OF HON. JERI MUOIO, MAYOR, WEST PALM BEACH, FLORIDA

Ms. MUOIO. Thank you, Senator. We are so happy to have you here in West Palm Beach. I just wanted to say that, in West Palm Beach, we say the words "climate change," and we say them frequently. We are very active. And I want to thank you for being here, first of all, and for holding this hearing.

I want to welcome our witnesses and, of course, welcome all the guests who are here this afternoon who have undertaken this.

In West Palm Beach, it has been very clear to us that climate change is occurring. We see it happening.

And I want to thank you, Senator Nelson, for very clearly explaining why we are seeing sea-level rise and why this is happening in our coastal regions.

What I would like to do is just take a couple minutes to talk about some of the things that we have done here in West Palm Beach.

We have had a sustainability action plan since 2008, and we have been working on that. We have these green bottles here that say, "Rethink Paradise." That is the title of our sustainability action plan. Penni Redford, who is the Manager of the Mayor's Office for Sustainability, she and I had the opportunity to go to Paris to attend the COP 21 meetings and to urge national leaders to commit to addressing climate change and making sure that there is a lowering of our temperature by a significant portion.

We have committed to that. I hope the United States will remain committed to it.

Here in West Palm Beach, we have committed to Net Zero by 2050. We have already begun to decrease our carbon emissions and have made a significant difference in that.

We have also pledged, by 2025, to turn over our fleet to non-fossil-fuel, and we are working on that diligently. It is not all that easy because there are not a lot of options out there for vehicles. We have a storm water master plan that we are working on that is going to make a significant difference in how we manage storm water. We are already seeing storm water intrusion on Flagler Drive, especially during king tide and when we have water events, rain events. It really will make a big difference.

Our city has committed to a Global Covenant of Mayors for Climate & Energy, the world's largest cooperative effort among mayors and city officials.

So we are very proud of that. And one of the things we are most proud of is that we were recently awarded a four-star rating by the Star Community Rating System. We are the only city in Florida to have achieved a four-star rating, and only a handful of cities in the country have achieved a five-star rating, so we are working on that diligently.

We have adopted new green building requirements, because we are going to have to address sea-level rise. We are talking about freeboarding. We are talking about alternative fuels as part of our Energy Secure Cities Coalition. We are moving forward with making sure that West Palm Beach is going to be a sustainable city as well as a resilient city.

So thank you for being here. We are very proud and honored that you chose to be here, and I am looking forward to hearing the testimony this afternoon.

Thank you.

Senator NELSON. Thank you, Madam Mayor.

And we are delighted to be here in Palm Beach County, and we have our County Mayor with us, Paulette Burdick, to share with us.

STATEMENT OF HON. PAULETTE BURDICK, MAYOR, PALM BEACH COUNTY, FLORIDA

Ms. BURDICK. Good afternoon, Senator. Welcome to Palm Beach County. As Mayor of the County, we are pleased and honored that you chose West Palm Beach, which is one of our cities in Palm Beach County, as your location to hold the U.S. Senate Commerce Committee field hearing on extreme weather and coastal flooding. We are honored to have our Congressman Deutch here. And it is always a delight to be with my mayor, Mayor Muoio.

Climate change has been characterized as the most profound threat to our national security and our country's continued prosperity. It does create instability in our economy through longer droughts and flooding. It harms our citizens' health through more frequent heat waves and air pollution. And it destabilizes our infrastructure through sea-level rise and extreme weather events.

As these conditions become our Nation's new norm, our collective response should be to strengthen critically important U.S. climate policies, including the Clean Power Plan and vehicle fuel efficiency standards, not proposing budget cuts to the EPA and crucial Federal programs like Energy Star.

The county is committed to improving the resilience of our communities both regionally and locally. As a charter member of the Southeast Florida Regional Climate Change Compact, a regional collaborative model to reduce carbon emissions, adapt to climate impacts, and build resilient communities, we strive to address the impacts of climate change, including sea-level rise, through regional collaboration and local action.

In the past 8 years, the climate compact has fostered close, cooperative relationships with Palm Beach, Broward, Miami-Dade, and Monroe counties, and our local municipalities.

Locally, Palm Beach County continues its effort to adapt to and mitigate the impacts of climate change. We are incorporating resilience and sustainability measures in our capital projects, curbing greenhouse gases through reducing soft costs for solar heating systems and tracking energy use, and by strengthening relationships with businesses and addressing economic benefits of resiliency in our community.

Politicizing science erodes one of the main hallmarks of our great country, the celebration of our ingenuity and the ability to transform insurmountable hurdles into advantageous opportunities. We urge the Senate Commerce Committee and its counterpart in the House to increase rather than eliminate local support, regional and national efforts to protect our businesses and our citizens from climate change, which continues to be documented and proven by the international community of scientists.

I call upon you to continue to provide the critical leadership needed on the global climate stage so the gains achieved over the past few years are not lost.

And on behalf of all of us here in Florida, thank you for your leadership and being with us here today.

Senator NELSON. Thank you, Mayor.

Okay, our distinguished panel, your written remarks will be included as part of the official record. What I would like you to do is to summarize your remarks and try to keep it within less than 5 minutes.

Dr. Ben Kirtman is the Director of the Center for Computational Science, and Climate and Environmental Hazards at the University of Miami—this is Dr. Kirtman over here—as well as the Director of the National Oceanographic and Atmospheric Administration's Cooperative Institute for Marine and Atmospheric Sciences. Dr. Kirtman is a professor in the Division of Meteorology and Physical Oceanography at UM's Rosenstiel School for Marine and Atmospheric Science, and is the Executive Editor of the scientific journal *Climate Dynamics.*

How do you have time for all this?

[Laughter.]

Senator NELSON. His research focuses on predicting climate change in the short and long term, and how much of that change can be attributed to humans.

Then we are going to hear from Dr. Leonard Berry. He is Professor Emeritus of Geosciences at Florida Atlantic, and he serves as a consultant to Coastal Risk Consulting, a Florida company that assesses climate risk to properties. He is also providing consulting services to communities who wish to relocate power infrastructure to avoid those risks.

Mr. Carl Hedde is the Senior Vice President and the Head of Risk Accumulation for Munich Reinsurance Company of America. Mr. Hedde oversees corporate accumulation issues, including use of catastrophic risk models, climate catastrophic risk consulting services, and portfolio management and optimization. Additionally, he manages a group of scientists that provide expert and research capabilities to Munich Reinsurance Company of America and its clients.

By the way, if you are confused, you do not hear a lot of insurance companies come out and chronicle this, though reinsurance companies do, because they see the long-term risk whereas the insurance companies are usually just setting their rates on the basis of 1 to 3 years in the future, not the reinsurance companies. That is why we have Mr. Hedde here today.

Then, Dr. Jennifer Jurado, she is the Chief Resiliency Officer as well as the Director of the Environmental Planning and Community Resilience Division of Broward County. Dr. Jurado serves as the county's primary representative coordinating with regency and agency partners—and we want you to tell us about the South Florida compact—and public and private stakeholders to advance the planning and infrastructure investments.

Dr. Jurado was on the President's Task Force on Climate Preparedness and was recognized in 2013 by the White House as a Champion of Change for her leadership on climate resilience. She was integral in drafting the compact, which she is going to tell us about.

So thank you all for being here. Let's just start in the order that I introduced you.

Dr. Kirtman.

STATEMENT OF BEN KIRTMAN, Ph.D., DIRECTOR, COOPERATIVE INSTITUTE FOR MARINE AND ATMOSPHERIC SCIENCES, AND DIRECTOR OF THE CENTER FOR COMPUTATIONAL SCIENCE, CLIMATE, AND ENVIRONMENTAL HAZARDS AT THE UNIVERSITY OF MIAMI

Dr. KIRTMAN. Senator Nelson, Congressman Deutch, distinguished Mayors, distinguished elected officials, thank you for this opportunity to share what I think are really important problems in climate change and climate variability. As a Floridian, I am particularly grateful for the Committee's focus on this matter as it hits very close to home for many of us.

First of all, as a scientist, my goal is to understand how the Earth system works and how to predict its evolution into the future. As a weather and climate scientist, it is my hope that policymakers will utilize the best available science to save lives, protect property, and enable economic opportunity, and secure our national defense.

What I want to do today is I want to tell you a little bit about the best available science that we have on climate change and how that affects Florida regionally. As part of that process, I want to have a mantra that sits in the background, and this is from the Intergovernmental Panel on Climate Change. That mantra is "multiple lines of evidence."

So I am going to show you a few figures, and they are going to be very overwhelming. That is on purpose. There are going to be tons and tons of curves, and I am not going to explain each individual curve. But I want to hit hard on this mantra of multiple lines of evidence.

Let's first start out with the fact that Senator Nelson has already given us, and that is that CO_2 and methane affect global atmospheric temperatures. That is a fact. Over the last 800,000 years, up until about 3,000 years, CO_2 levels have oscillated between 180 parts per million and 280 parts per million—180 to 280. Those oscillations took on the order of 10,000 to 40,000 years.

The current levels of CO_2 are 405. That jump from 280 to 405 took less than 200 years—less than 200 years. Something is out of balance.

What I show here on the panel is our instrument record, our modern instrument record, and you can see there are a number of different fields there. They have different curves and different data sets. These are disparate scientists using different instruments to collect this data and to analyze the data. And it is different fields. One might be Arctic sea ice in the summer. One might be sea surface temperatures. One is sea-level rise.

Multiple lines of evidence all point to the same thing, all point to the same thing, that our climate system is out of balance, and it has warmed, 100 percent confidence, since the 1950s.

I want to pause a moment to talk about sea level because this is important. This is really important. That top panel shows sea-level rise estimates from the previous 2,000 years up until about 1900, or maybe 1700—2,000 years.

What you notice in that top panel is there is no change in sea level. The sea level is remarkably stable. When you look more recently in the other two panels to the left, what do you see? You see over the modern instrument record, a rapid increase in sea level globally. That is due to human activities.

The panel on the right, that is flooding in Miami Beach. These are based on insurance claims and anecdotal evidence but also data. What you can see is the rate of flooding has steadily increased over that record.

So these are clear, unequivocal indicators, 95 percent to 100 percent certainty—scientists are always a little bit conservative—95 percent to 100 percent certainty that the climate system has warmed and the sea level has risen since the 1950s, and that is due to human activities.

How do we know it is due to human activities? You can analyze the data very carefully. What this is is a very, very careful analysis of the global temperature records. And you can see we can contribute it to El Nino, which we just had. We can attribute it to sunspot cycles. We can attribute it to all kinds of things. Volcanoes.

The one thing that jumps off the page is the one in the black oval. That is human activity. The bulk of the temperature rise we have seen since the 1950s is due to human activities.

We do the same thing with our model that is in the right-side panel. We take out the volcanoes. We take out the CO_2. We are unable to reproduce the temperature record in the 20th century without the anthropogenic greenhouse gases.

Again, multiple lines of evidence all pointing to the same thing.

The last slide, the regional change is difficult, it is a very difficult problem, and that is because the natural variability is large. It requires very careful calculations. But the panels on the right are showing the near-term future. We have every reason to believe that the current trends that we are seeing are going to continue.

In fact, there is compounding evidence that the current trends are going to continue. There is no credible science whatsoever that the trends we see today are going to reverse themselves.

So even if you want to be skeptical about human activities affecting climate change, for the next 25 to 50 years, there is no evidence that those trends will reverse.

Thank you very much.

[The prepared statement of Dr. Kirtman follows:]

PREPARED STATEMENT OF BEN KIRTMAN, PH.D., PROFESSOR, UNIVERSITY OF MIAMI— ROSENSTIEL SCHOOL FOR MARINE AND ATMOSPHERIC SCIENCE; DIRECTOR, COOPERATIVE INSTITUTE FOR MARINE AND ATMOSPHERIC STUDIES; PROGRAM DIRECTOR FOR CLIMATE AND ENVIRONMENTAL HAZARDS, CENTER FOR COMPUTATIONAL SCIENCE

Mr. Chairman, Ranking Member Nelson, distinguished members of the U.S. Senate Committee on Commerce, Science, and Transportation and members of the House of Representatives, thank you for this opportunity to come before you today and share my thoughts regarding climate variability and climate change, and how this affects Florida.

I've been a climate scientist for about 25 years, having received my Ph.D. in 1992. Ten years ago, I joined the University of Miami Rosenstiel School of Marine and Atmospheric Science as a Professor of meteorology and physical oceanography and in 2016 was appointed as the Director of the Cooperative Institute of Marine and Atmospheric Studies. I use complex earth system models and the most sophisticated supercomputers throughout the United States to investigate the predictability of the climate system on time scales from days-to-decades.

I served as a coordinating lead author for the Intergovernmental Panel on Climate Change (IPCC) working group one—the Scientific Basis and have chaired several national and international scientific panels and working groups. I'm an Executive Editor of *Climate Dynamics* and an Associate Editor of the *American Geophysical Union Journal of Geophysical Research* (Atmospheres). I have received research grants from the National Science Foundation, Department of Energy, NOAA, NASA, and the Office of Naval Research, and I lead the North American Multi-Model Ensemble Prediction (NMME) Experiment. I'm the author and/or coauthor of over 120 peer reviewed papers focused on understanding and predicting climate variability on time scales from days to decades.

And as a Floridian, I am grateful for the Committee's focus on a matter that hits very close to home for many of us in this room.

First and foremost, as a scientist my goal is to understand how the earth system works and how to predict its evolution into the future. As weather and climate scientists, it is our hope that policy makers will be able to utilize the best available science to help: (1) save lives, (2) protect property, (3) enable economic opportunity and (4) secure our national defense.

My testimony will summarize the current state-of-the-science in climate variability and change on a global scale, and how these global drivers affect the local Florida environment. The over-arching key points are summarized below, and the remaining text goes into further detail with data and figures. Much of the material included here is from the Intergovernmental Panel on Climate Change (IPCC) 5th Assessment Report (AR5, Stocker *et al.*, 2013; Kirtman *et al.*, 2013), which assesses our current scientific understanding of climate change. It is important to understand that any robust conclusions in the IPCC assessment report require: (i) multiple disparate lines of evidence and (ii) quantitative estimates of uncertainty. This assessment process summarizes the best available science.

The Science: Global Climate Drivers of Regional Change

(i) CO_2 levels in the atmosphere affect global temperatures.

(ii) During the last 800,000 years (excluding the modern era; 1900-present), CO_2 levels in the atmosphere have ranged from about 180 parts per million by volume (ppmv) to about 280 ppmv. The oscillations were between 180 and 280 ppmv; these changes took approximately 10,000 to 40,000 years to occur. Current CO_2 levels are about 405 ppmv and the increase from 280 to 405 ppmv took less than 150 years (see Fig. 1). This rapid increase in CO_2 is unprecedented in any observational estimate.

(iii) Since the 1950s the climate system has warmed and it is 100 percent unequivocal (see Fig. 2). There are robust multiple lines of evidence—multiple studies that involve different observational instruments that measure different components of the climate system—that support this conclusion.

(iv) The bulk of the warming since the 1950s is extremely likely (95–100 percent certainty) due to human activities (*i.e.*, increases in CO_2 levels associated with the consumption of fossil fuels; see Fig. 3 and Fig. 4).

(v) Given its importance in Florida, sea level merits special attention. Paleo sea level data from the last 3000 years, until approximately 1900, has been remarkably stable; there has been little change in the global mean. However, since about 1900 global mean sea level has steadily risen consistent with the warming seen (Fig. 5).

(vi) Regional climate changes are more difficult to assess. This is because the natural variability tends to be larger on the local scale, and this makes it more challenging to isolate the anthropogenic signal. Nevertheless, regional changes in temperature through out much of the U.S. show a pronounced warming trend (see Fig. 6).

(vii) There is evidence that at regional scales along the eastern U.S., and in Florida in particular, the sea level rise is accelerating (see Fig. 7).

(viii) There is *no* compelling scientific evidence that any of the trends that we currently see are going to reverse themselves. There is, however, compelling evidence that the current trends will continue for at least the next 25 years, and there is even some evidence that particular trends may accelerate. Even if one is skeptical that human activities are the cause of these trends, there is a clear local need to protect lives and property, and ensure economic opportunity in response to changes we see today. Robust, well-calibrated, scientifically based predictions of the next 25 years and beyond (see Fig. 8) are the first step in developing effective adaptation strategies and to capitalize on the associated economic opportunities.

(ix) Florida is well positioned to respond to the challenges and opportunities associated with climate change. The academic community has established the Florida Climate Institute (FCI; *https://floridaclimateinstitute.org*). The Florida Climate Institute (FCI) fosters interdisciplinary research, education, and extension to: Improve our understanding and the impact of climate variability, climate change, and sea level rise on the economy, ecosystems, and human-built systems; Develop technologies and information for creating opportunities and policies that reduce economic and environmental risks; and Engage society in research, extension and education programs for enhancing adaptive capacity and responses to associated climatic risks. We collaborate with the local, state, and Federal Government to address our most pressing adaptation problems.

(x) The process of challenging the conventional wisdom is a critical component of how robust science progresses. We should always be respectful of differing perspectives, accounting for new information and ideas and then test them through the scientific method. This is how science works, this is how we find fact. When it comes to policy, I would just ask that policy makers take into account the best available science. When it comes to climate change, the scientific consensus is not cavalier, it is prudent and conservative, and is the best available science.

Basic Global Climate Change

Figure 1 shows 800,000 years of CO_2 and temperature from ice core records from Vostok, Antarctica. The temperature near the South Pole has varied by as much as 20°F (11°C) during the past 800,000 years. The cyclical pattern of temperature variations constitutes the ice age/interglacial cycles. During these cycles, changes in carbon dioxide concentrations (in purple) track closely with changes in temperature (in blue), with CO_2 lagging behind temperature changes. Because it takes a while for snow to compress into ice, ice core data are not yet available much beyond the 18th century at most locations. However, atmospheric carbon dioxide levels, as measured in air, are higher today than at any time during the past 800,000 years. Source: National Research Council (*https://nas-sites.org/americasclimatechoices/more resources-on-climate-change/climate-change-lines-of-evidence-booklet/evidence-impacts-andchoices-figure-gallery/figure-14/*).

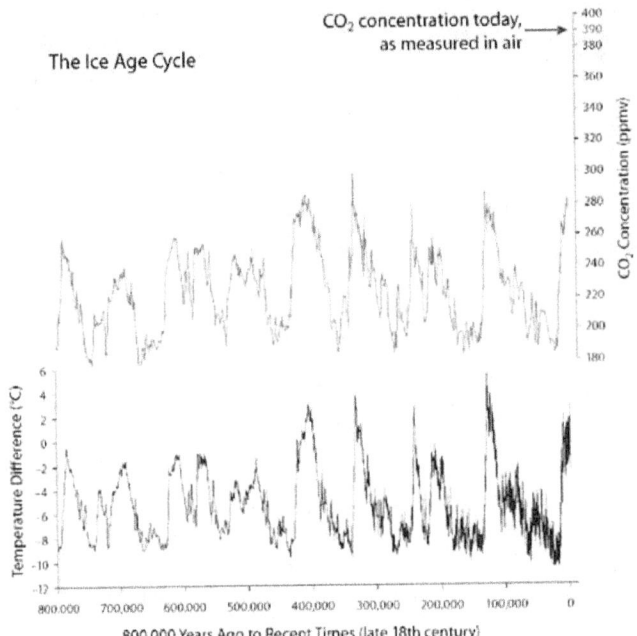

Source 1 for top image: Lüthi, D., M. Le Floch, B. Bereiter, T. Blunier, J.-M. Barnola, U. Siegenthaler, D. Raynaud, J. Jouzel, H. Fischer, K. Kawamura, and T. F. Stocker. 2008. High-resolution carbon dioxide concentration record 650,000–800,000 years before present. Nature 453(7193):379–382, doi: 10.1038/nature06949.

Source 2 for bottom image: Jouzel, J., V. Masson-Delmotte, O. Cattani, G. Dreyfus, S. Falourd, G. Hoffmann, B. Minster, J. Nouet, J. M. Barnola, J. Chappellaz, H. Fischer, J. C. Gallet, S. Johnson, M. Leuenberger, L. Loulergue, D. Luethi, H. Oerter, F. Parrenin, G. Raisbeck, D. Raynaud, A. Schilt, J. Schwander, E. Selmo, R. Souchez, R. Spahni, B. Stauffer, J. P. Steffensen, B. Stenni, T. F. Stocker, J. L. Tison, M. Werner, and E. W. Wolff. 2007. Orbital and millennial Antarctic climate variability over the past 800,000 years. Science 317(5839):793–797.

One of the top level conclusions of the IPCC AR5 is that the since the 19th century the climate system has warmed. This conclusion is based on multiple lines of evidence from many different data sets that have been collected using different instruments. All of these data sets, whether they are ocean, land, or sea-ice measurements point to one unequivocal conclusion—the world has warmed. Figure 2 summarizes the results from many several of these different data sets. For example, there are four different data sets used to estimate global land surface temperature changes, and they all indicate a warming of about 2.5°F. There are six different data set used to estimate global sea level, and again they all agree in the upward trend. Summer arctic sea-ice extent is estimated using six different data sets, and they all indicate the same downward trend.

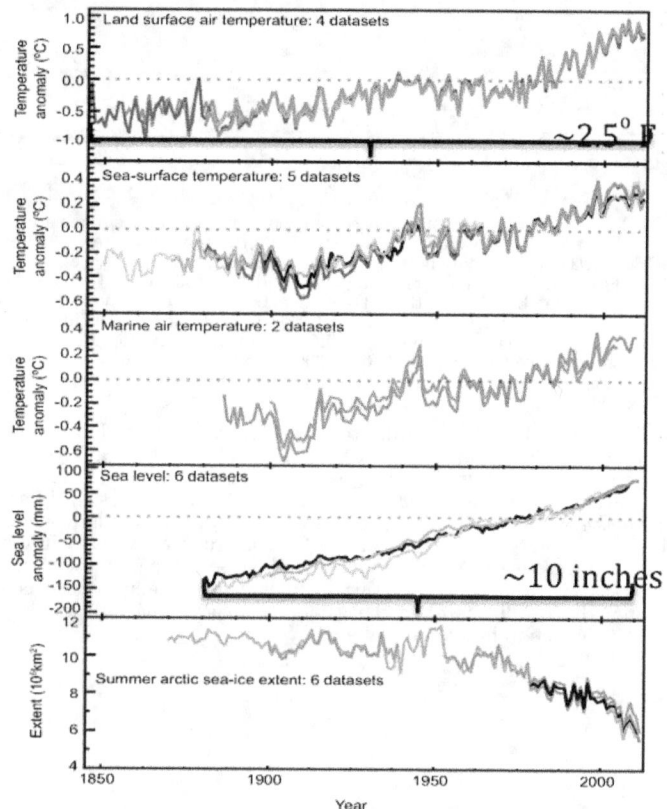

Figure 2: Multiple independent indicators of a changing global climate. Each line represents an independently derived estimate of change in the climate element. In each panel all data sets have been normalized to a common period of record. Figure take from IPCC AR5 and a full detailing of the data sources is given in Stocker *et al.*, (2013, supplementary material).

Perhaps the most important question that needs to be addressed is how do we know the trends seen in Fig. 2 are due to human activities. There are two typical approaches. The first is referred to detection and attribution studies (Bindoff *et al.*, 2013). Figure 3 summarize a classic detection and attribution study based on observational estimates of global mean surface temperatures. The time series analysis separates the global mean temperature changes due to: El Niño (panel b), volcanoes (panel c), solar output (panel d), and other modes of climate variability like the AMO (panel f). The global mean temperature changes associated with the changes in greenhouse gases such a CO_2 are shown in panel e, and demonstrate that it is extremely likely (95–100 percent) that the bulk of the warming since the 1950s is due to human activities.

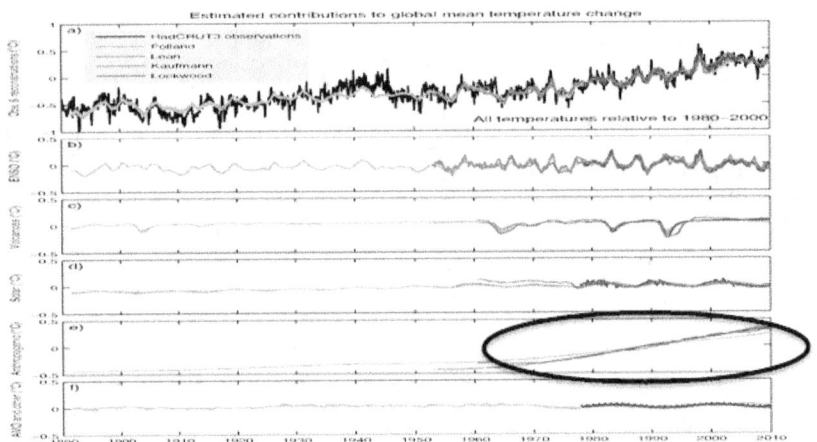

Figure 3: (Top) The variations of the observed global mean surface temperature (GMST) anomaly from Hadley Centre/Climatic Research Unit gridded surface temperature data set version 3 (HadCRUT3, black line) and the best multivariate fits using the method of Lean (red line), Lockwood (pink line), Folland (green line) and Kaufmann (blue line). (Below) The contributions to the fit from (a) El Niño-Southern Oscillation (ENSO), (b) volcanoes, (c) solar forcing, (d) anthropogenic forcing and (e) other factors (Atlantic Multi-decadal Oscillation (AMO) for Folland and a 17.5-year cycle, semi-annual oscillation (SAO), and Arctic Oscillation (AO) from Lean). (From Lockwood (2008), Lean and Rind (2009), Folland et al. (2013) and Kaufmann et al. (2011), as summarized in Imbers et al. (2013).) See Figure 10.6 in Bindoff et al. (2013) for references and details.

The second approach for attributing the observed warming to human activities is based on climate model simulations. Again, as with the data analysis shown in Fig. 3, the climate models used in the assessment of the climate of the 20th century have been developed and validated by different modeling centers in different countries around the world—multiple lines of evidence supporting the conclusion. The approach is to simulate the climate of the 20th century with and without the anthropogenic changes in CO_2. The results and then be compared with the observed temperature record. An example of this for global mean temperature is shown in Fig. 4. Again, the results point to the same conclusion—the bulk of the warming since the 1950s is due to human activities.

Land and ocean surface

Figure 5: Comparison of observed (black curve) and multi-model simulated global mean temperature with natural and anthropogenic forcing (pink swath) and just natural forcing (blue swath). The width of the swaths correspond to the 5 percent–95 percent range from the multi-model ensemble simulations. Figure adapted from Stocker *et al.*, 2013.

Sea level rise associated with climate change is of particular importance to Florida. Here we show results from Church *et al.*, (2013) which includes a detailed analysis of paleo and historical estimates of global sea level and more recent modern instrument records. The results further underscore the unequivocal conclusion that human activities are leading to profound changes in the climate system.

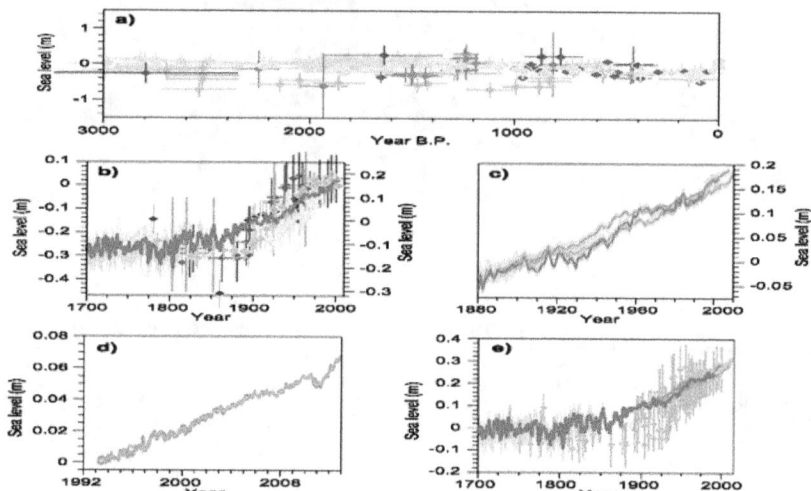

Figure 5: a) Paleo sea level data for the last 3,000 years from Northern and Southern Hemisphere sites. The effects of glacial isostatic adjustment (GIA) have been removed from these records. Light green = Iceland (Gehrels *et al.*, 2006), purple = Nova Scotia (Gehrels *et al.*, 2005), bright blue = Connecticut (Donnelly *et al.*, 2004), blue = Nova Scotia (Gehrels *et al.*, 2005), red = United Kingdom (Gehrels *et al.*, 2011), green = North Carolina (Kemp *et al.*, 2011), brown = New Zealand (Gehrels *et al.*, 2008), grey = mid-Pacific Ocean (Woodroffe *et al.*, 2012). (b) Paleo sea level data from salt marshes since 1700 from Northern and Southern Hemisphere sites compared to sea level reconstruction from tide gauges (blue time series with uncertainty) (Jevrejeva

et al., 2008). The effects of GIA have been removed from these records by subtracting the long-term trend (Gehrels and Woodworth, 2013). Ordinate axis on the left corresponds to the paleo sea level data. Ordinate axis on the right corresponds to tide gauge data. Green and light green = North Carolina (Kemp et al., 2011), orange = Iceland (Gehrels et al., 2006), purple = New Zealand (Gehrels et al., 2008), dark green = Tasmania (Gehrels et al., 2012), brown = Nova Scotia (Gehrels et al., 2005). (c) Yearly average global mean sea level (GMSL) reconstructed from tide gauges by three different approaches. Orange from Church and White (2011), blue from Jevrejeva et al., (2008), green from Ray and Douglas (2011) (see Section 3.7). (d) Altimetry data sets from ve groups (University of Colorado (CU), National Oceanic and Atmospheric Administration (NOAA), Goddard Space Flight Centre (GSFC), Archiving, Validation and Interpretation of Satellite Oceanographic (AVISO), Commonwealth Scientific and Industrial Research Organisation (CSIRO)) with mean of the ve shown as bright blue line (see Section 3.7). (e) Comparison of the paleo data from salt marshes (purple symbols, from (b)), with tide gauge and altimetry data sets (same line colours as in (c) and (d)). All paleo data were shifted by mean of 1700–1850 derived from the Sand Point, North Carolina data. The Jevrejeva et al., (2008) tide gauge data were shifted by their mean for 1700–1850; other two tide gauge data sets were shifted by the same amount. The altimeter time series has been shifted vertically upwards so that their mean value over the 1993–2007 period aligns with the mean value of the average of all three tide gauge time series over the same period. References and details in Church et al., 2013.

Regional Climate Change

Regional climate changes are more difficult to assess. This is because the natural variability tends to be larger on the local scale, and this makes it more challenging to isolate the anthropogenic signal. Nevertheless, regional changes in temperature thought much of the U.S. show a pronounced warming trend (see Fig. 6).

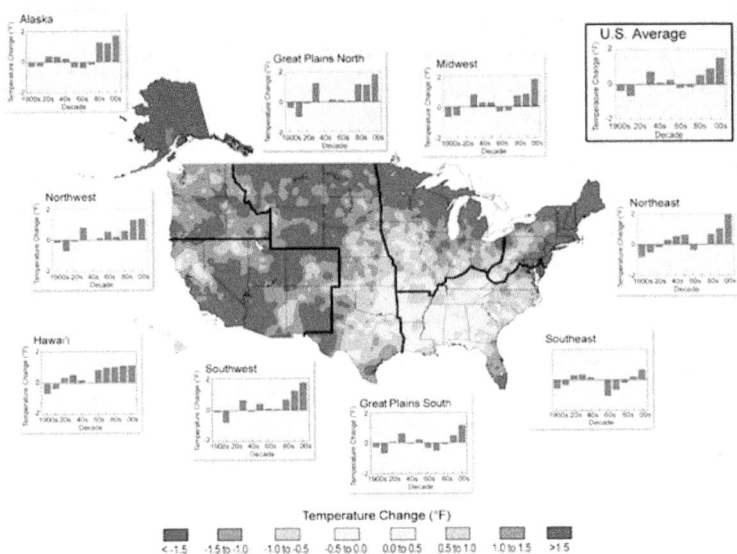

Figure 6: The colors on the map show temperature changes over the past 22 years (1991–2012) compared to the 1901–1960 average, and compared to the 1951–1980 average for Alaska and Hawai'i. The bars on the graphs show the average temperature changes by decade for 1901–2012 (relative to the 1901–1960 average) for each region. The far right bar in each graph (2000s decade) includes 2011 and 2012. The period from 2001 to 2012 was warmer than any previous decade in every region. (Figure source: NOAA NCDC / CICS–NC). Figure taken from Melillo et al., 2014.

There is evidence that sea level rise along the eastern seaboard of the U.S. is accelerating (Fig. 7 below). The factors for the acceleration are not well understood but could be due to changes in ocean circulation associated with global warming, Greenland ice melt also associated with global warming or even land subsidence.

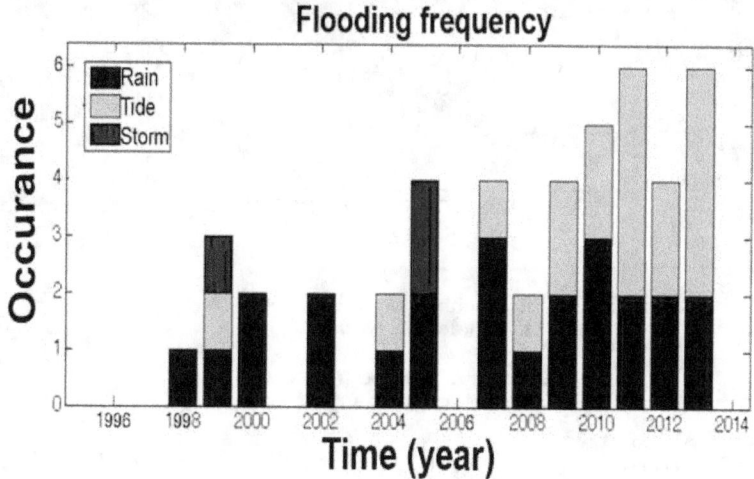

Figure 7: Flooding frequency in Miami Beach. Figure adapted from Wdowinski *et al.*, (2016)

Finally, there is *no* compelling scientific evidence that any of the trends that we currently see are going to naturally? reverse themselves. There is, however, compelling evidence that the current trends will continue for at least the next 25 years, and there is even some evidence that particular trends (regional sea level) may accelerate (see discussion of Fig. 7).

Predicting the Future

Even if one is skeptical that human activities are the cause of these trends, there is a clear local need to protect lives and property, and ensure economic opportunity in response to changes we see today. Robust well-calibrated scientifically based predictions of the next 25-years (and beyond) are the first stop in developing effective adaptation strategies and to capitalize on the associated economic opportunities.

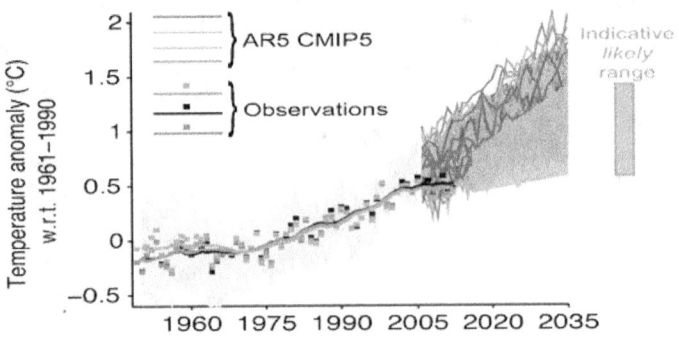

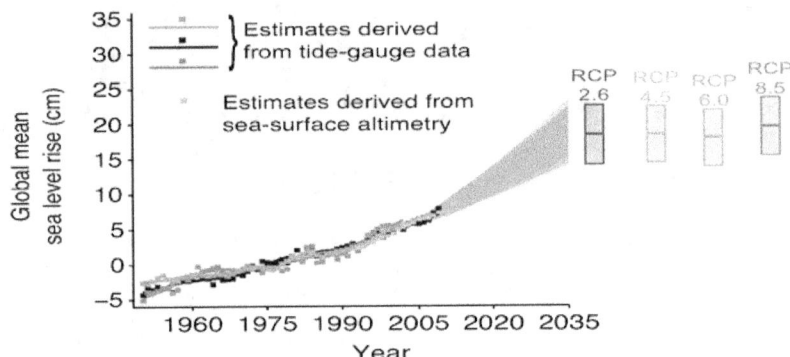

Figure 8: Projected changes in global temperature and sea level from IPCC AR5. See Stocker *et al.*, 2013 for details.

References

Bindoff, N.L., P.A. Stott, K.M. AchutaRao, M.R. Allen, N. Gillett, D. Gutzler, K. Hansingo, G. Hegerl, Y. Hu, S. Jain, I.I. Mokhov, J. Overland, J. Perlwitz, R. Sebbari and X. Zhang, 2013: Detection and Attribution of Climate Change: from Global to Regional. In: Climate Change 2013: *The Physical Science Basis. Contribution of Working Group I to the Fifth Assessment Report of the Intergovernmental Panel on Climate Change* [Stocker, T.F., D. Qin, G.-K. Plattner, M. Tignor, S.K. Allen, J. Boschung, A. Nauels, Y. Xia, V. Bex and P.M. Midgley (eds.)]. Cambridge University Press, Cambridge, United Kingdom and New York, NY, USA.

Church, J.A., P.U. Clark, A. Cazenave, J.M. Gregory, S. Jevrejeva, A. Levermann, M.A. Merri eld, G.A. Milne, R.S. Nerem, P.D. Nunn, A.J. Payne, W.T. Pfeffer, D. Stammer and A.S. Unnikrishnan, 2013: Sea Level Change. In: *Climate Change 2013: The Physical Science Basis. Contribution of Working Group I to the Fifth Assessment Report of the Intergovernmental Panel on Climate Change* [Stocker, T.F., D. Qin, G.-K. Plattner, M. Tignor, S.K. Allen, J. Boschung, A. Nauels, Y. Xia, V. Bex and P.M. Midgley (eds.)]. Cambridge University Press, Cambridge, United Kingdom and New York, NY, USA.

Collins, M., R. Knutti, J. Arblaster, J.-L. Dufresne, T. Fichefet, P. Friedlingstein, X. Gao, W.J. Gutowski, T. Johns, G. Krinner, M. Shongwe, C. Tebaldi, A.J. Weaver and M. Wehner, 2013: Long-term Climate Change: Projections, Commitments and Irreversibility. In: *Climate Change 2013: The Physical Science Basis. Contribution of Working Group I to the Fifth Assessment Report of the Intergovernmental Panel on Climate Change* [Stocker, T.F., D. Qin, G.-K. Plattner, M. Tignor, S.K. Allen, J. Boschung, A. Nauels, Y. Xia, V. Bex and P.M. Midgley (eds.)]. Cambridge University Press, Cambridge, United Kingdom and New York, NY, USA.

Kirtman, B., S.B. Power, J.A. Adedoyin, G.J. Boer, R. Bojariu, I. Camilloni, F.J. Doblas-Reyes, A.M. Fiore, M. Kimoto, G.A. Meehl, M. Prather, A. Sarr, C. Schär, R. Sutton, G.J. van Oldenborgh, G. Vecchi and H.J. Wang, 2013: Near-term Climate Change: Projections and redictability. In: Climate Change 2013: The Physical

Science Basis. Contribution of Working Group I to the Fifth Assessment Report of the Intergovernmental Panel on Climate Change [Stocker, T.F., D. Qin, G.-K. Plattner, M. Tignor, S.K. Allen, J. Boschung, A. Nauels, Y. Xia, V. Bex and P.M. Midgley (eds.)]. Cambridge University Press, Cambridge, United Kingdom and New York, NY, USA.

Melillo, Jerry M., Terese (T.C.) Richmond, and Gary W. Yohe, Eds., 2014: *Climate Change Impacts in the United States: The Third National Climate Assessment.* U.S. Global Change Research Program, 841 pp. doi:10.7930/J0Z31WJ2.

Stocker, T.F., D. Qin, G.-K. Plattner, L.V. Alexander, S.K. Allen, N.L. Bindoff, F.-M. Bréon, J.A. Church, U. Cubasch, S. Emori, P. Forster, P. Friedlingstein, N. Gillett, J.M. Gregory, D.L. Hartmann, E. Jansen, B. Kirtman, R. Knutti, K. Krishna Kumar, P. Lemke, J. Marotzke, V. Masson-Delmotte, G.A. Meehl, I.I. Mokhov, S. Piao, V. Ramaswamy, D. Randall, M. Rhein, M. Rojas, C. Sabine, D. Shindell, L.D. Talley, D.G. Vaughan and S.-P. Xie, 2013: Technical Summary. In: *Climate Change 2013: The Physical Science Basis. Contribution of Working Group I to the Fifth Assessment Report of the Intergovernmental Panel on Climate Change* [Stocker, T.F., D. Qin, G.-K. Plattner, M. Tignor, S.K. Allen, J. Boschung, A. Nauels, Y. Xia, V. Bex and P.M. Midgley (eds.)]. Cambridge University Press, Cambridge, United Kingdom and New York, NY, USA.

Wdowinski, S. R. Bray, B. P. Kirtman, Z. Wu, 2016: Increasing flooding hazard in coastal communities due to raising sea level: Case study of Miami Beach, Florida. *Ocean and Coastal Management,* **126,** 1–8.

Senator NELSON. Thank you very much, Dr. Kirtman.
Dr. Berry.

STATEMENT OF LEONARD "LEN" BERRY, Ph.D., EMERITUS PROFESSOR, GEOSCIENCES, FLORIDA ATLANTIC UNIVERSITY, AND VICE PRESIDENT, GOVERNMENT PROGRAMS, COASTAL RISK CONSULTING, LLC

Dr. BERRY. Senator, Congressman, Mayors, 5 years ago, almost to the day, I testified before the Senate Committee on Natural Resources in Washington, and I started with a joke. I said, "the sky is not falling, but the seas are rising." It got their attention. Now I can tell you the seas are still rising, and I wonder sometimes if the skies are falling.

What I thought I would try to do today was really talk about what has happened in those 5 years in relation to sea-level rise and our understanding.

On the science side, we have all the stuff that Ben has talked about, but we have also understood the processes better. We know more about the Antarctic and what is happening there. We know more about the role of the Gulf Stream in our local sea-level rise. We did not know a lot of that 5 years ago. We know it now. And as Ben says, the percent of certainty goes up from 95 beyond.

We know a lot more, and we are a lot more scared, because if I look at the new information coming up year by year, every time it gets a little bit worse. It is not inference. It is what data comes out, what information comes out.

In my written testimony, I emphasize, and we can emphasize more, that the information provided by NOAA, NASA, and all of our other agencies are vital for our understanding of what is happening now and for our understanding of what is going to happen in the near and more distant future.

Also, the data is vital for mayors, Congressmen, businesses. The business I co-founded relies on NASA data—LIDAR, sea-level tides, and so on. Information is the key to our understanding. And reliable, credible information is what we have to rely on.

Our appreciation of risk has changed over 5 years partly because of all of the stuff I am talking about but also, as you point out—you stole my thunder—it is about experience. We experience flooding. We spend $100 million a year in Miami Beach on pumps. We do all of this stuff, and it is new.

And so as our appreciation of risk has changed, our response has changed. We have sustainability officers—one great one here. A lot of counties, cities, now have sustainability officers, because they need people who understand a broad range of change and are willing to manage it in a relatively uncertain future.

I think our responses have changed. We were not spending $100 million anywhere on pumps 5 years ago. But king tides have convinced us that the problem is not in the future. It is now. Adaptation action scenarios have been designated, mainly in Broward County, and people in those areas have special legal issues and special responses.

Investments in people and technology are protecting more than the buildings and the infrastructure. They are protecting, as you point out, the economy. Our economy depends on our coastal areas being vital. And without responding to our sea-level rise issues, those coastal areas will be in trouble.

On the other hand, there are opportunities there that we have in using our technology to be innovative, create new ideas, and create new jobs.

Homeowners have a responsibility too. The company I am working with focuses on individual homeowners. We can provide them with information on what they need to do. They can match what the communities are going to do and become a stable community where the central organization is not doing one thing and the homeowners are trying to catch up. Homeowners need to take adaptation action, and they need data to do it.

This may sound good as a regional response, and we feel very proud of it, but it pales in significance beyond what needs to be done.

As you pointed out and as we all know, great things done at the regional level have to be matched at the State level and at the Federal level. We need a coherent policy that deals with all of those levels at one time, and I hope that we can find ways that this Committee can do them.

[The prepared statement of Dr. Berry follows:]

PREPARED STATEMENT OF LEONARD "LEN" BERRY, PH.D., EMERITUS PROFESSOR, GEOSCIENCES, FLORIDA ATLANTIC UNIVERSITY, AND VICE PRESIDENT, GOVERNMENT PROGRAMS, COASTAL RISK CONSULTING, LLC

I am Dr. Leonard Berry, founder and former director of the Florida Center for Environmental Studies at Florida Atlantic University (FAU), now Emeritus professor of geosciences at FAU. I am also the co-founder and Vice President of Coastal Risk Consulting, LLC, a start-up technology company providing affordable flood vulnerability and risk assessments, located in Plantation, Broward County, Florida.

As a scientist and businessman, and as a resident, I want to emphasize that sea level rise is a critical issue in South Florida—not only in Miami, which gets most of the press—but for the whole region, including here in Palm Beach County. The issue requires our scientific, legislative and legal attention; and our current investment in the amount of several billions of dollars to be spent over the next few years alone.

As the topics of this hearing suggest, dealing with the current and future threat of sea level rise requires a combination of the best available science, based on credible and ongoing global data, and assessment of that data with respect to conditions on the ground—from communities to individual properties—to determine degrees of risk now, and in the future.

But the important part is action across all parts of society, from government on the local to national levels, in the business community, and in civil society. While we have seen responses from many parts of our community in South Florida, the action is far from universal.

My colleague Dr. Ben Kirtman will address the scientific progress in some detail, but his work and that of others depend on the continued information flow from the Federal Government. I have attached a letter to the President from many Florida scientists that underscores the importance of science and information.

It can't be overemphasized—continued flow of global data is vital to our understanding of the science of sea level rise and the actions we take form that science. As the 2016 Sea-Level Rise Summit (*http://www.ces.fau.edu/arctic-florida/*) highlighted, the fate of Florida depends on what happens in the Arctic and Greenland. And over a longer timeframe, what happens in Antarctica is important too. There are large questions and uncertainties regarding the melt rate and effects of ice on land. We must continue to learn more and reduce these uncertainties.

South Florida and the state as a whole have long experience of dealing with risk in the form of major storms or hurricanes. Extreme weather is predicted—we prepare for it, and if it strikes, we respond and rebuild.

Risk from sea level rise is different however; steady increase in the number of sunny-day flooding events is more insidious, though sometimes, like during Sandy, the level of risk does not get fully exposed until an storm like no other.

In various ways the region has begun to assess the level of risk and to respond. The obvious threat is physical damage to infrastructure, but we now recognize that the economy is also at risk unless remedial actions are taken. The Southeast Florida Regional Climate Compact, an important partnership now supported scientifically by the statewide Florida Climate Institute, has begun laying out the response to these challenges. Action has been occurring a slower pace, albeit faster than almost anywhere else in the United States.

One of the Compact's activities is the Resilient Redesign series, where sample areas are re-planned to thrive in a future sea level situation. The Resilient Redesign concepts for part of Delray Beach are appended as examples.

Adaptation to sea level rise however, has both top-down and bottom-up components. Government on all levels will have a responsibility to assist, and individual homeowners have obligations as well.

In the last five years, we have observed case studies on sea level rise adaptation. The City of Miami Beach is on the forefront—raising streets, communicating directly with residents, and having the difficult, but necessary conversations about sea level rise. Recently, the Cities of Miami and Miami Beach with Miami-Dade County were honored by being accepted as part of the Rockefeller Foundation's 100 Resilience Cities program, further elevating the conversation and creating a cross-government organization to do so. These issues will not be solved by one level of government, but will require crosscutting and boundary-breaking ideas and actions. Dr. Jennifer Jurado, the Chief Climate Resilience Officer for Broward County, can more knowledgably speak to local government adaptation actions.

Private homeowners have a responsibility as well. Individual residents are going to have to take adaption actions to protect their homes and assets. In order to take necessary actions, homeowners need accurate and trustworthy actionable intelligence. This is where innovation and technology intersect with sea level rise.

After I retired from Florida Atlantic University, I co-founded a company based in South Florida called Coastal Risk Consulting. The company's mission is to help coastal residents in the United States and around the world prepare for sea level rise and coastal flooding. It has become my mission to use the best available science and distill that down to the individual property level to create a communication tool personalized to a homeowner. Coastal Risk forecasts the numbers of fair-weather flood days an individual property owner will see for the next 30 years. This is all presented in an adaptation framework, proposing adaptation steps based on the number of fair-weather flood days.

Homeowners can use this data to make the decision to adapt now, plan on adapting in the future, or decide that adaption will not be necessary for the foreseeable future. The important part of the decision making process is that it is a well-informed decision about personal adaptation. As cities and towns begin to plan for adaption, homeowner must as well. This is how we will create robust and resilient communities.

Sea level rise is a complex, multi-faceted issue that will require the best minds from all fields of study and industries to come together.

For example, in addition to protecting their assets through adaptation, homeowners protect their homes through insurance. As Congress begins to reauthorize the National Flood Insurance Program, we must also consider the effects of sea level rise. Big questions remain for the program—What will flood insurance look like in the future? When homes begin to flood due to fair-weather flooding associated with sea level rise, will that be an insured loss? What will happen to flood insurance in those areas that are now more susceptible to coastal inundation due to sea level rise?

Sea level rise will also raise new and complex environmental issues. How will we ensure that traditional septic tanks will not contaminate surrounding areas? What other issues will communities need to consider as salt water begins to interact with the built environment?

In addition, South Florida has its own set of complex issues. The limestone base of our community is basically a porous rock. So as sea level rises, the water table will be pushed closer to the surface. Communities close to sea level, even far inland, will see water seep through the surface creating puddles at low lying points. Properly communicating this risk is imperative, as sea level rise is currently considered solely an issue for communities that abut the coastline.

While sea level rise is acknowledged as a South Florida concern, it is becoming an issue for every coastal community in the world. Boston, Norfolk, Santa Monica, Charleston, New Orleans, and even our Nation's capital, Washington, DC, are beginning to see clear signs that the seas are rising. We want all of our cities to be resilient.

Adaptation to sea level rise and resiliency go hand in hand. We cannot create resilient communities along our coastlines if we do not begin the adaptation conversation now, and in order to have a conversation, we must have the best available data. States, local governments, and companies rely on this data from the Federal Government and we'll need it even more going forward.

Attachments: Letter to the President, dated March 13, 2017, Resilient Redesign notes and presentation excerpt

ATTACHMENTS

March 13, 2017

Dear President Trump:

American scientists have historically been at the forefront of scientific discoveries and innovation. America should invest heavily in our effort to understand our home planet and be aware of how physical changes will impact industry and society.

Two agencies in your Administration support such studies: the National Oceanic and Atmospheric Administration (NOAA) and the National Aeronautics and Space Administration (NASA). These agencies provide the knowledge that helps our crops grow stronger, our fisheries thrive, and our tourism flourish.

NOAA and NASA are crucial to our safety and our ability to predict extreme weather. Not only do NASA and NOAA Earth-observing programs create jobs, but they also have a huge impact on our economy, and ultimately generate the technologies that make America great.

Yet right now, competition from China and Europe could eclipse our programs. We need strong Earth science programs so we are able to prepare a better future.

NASA and NOAA Earth science programs monitor and understand changes in our water resources and our soil. They track the conditions that affect the food and medicines we get from the oceans. These conditions impact agriculture, our military, businesses, and major industries. It is imperative to support programs that explore our planet – at NOAA and NASA and across the government.

The work of NOAA and NASA is vital to life on Earth and must be continued.

As you turn your attention to new heads for these agencies, consider the critical role that they play in helping us survive and thrive on Earth.

As you ponder your selection, we encourage you to:

Continue Earth Science Research
America must continue funding both NOAA's environmental satellites and NASA's new technologies to measure the vital signs of our planet.

NASA and NOAA's work capture the history and the present state of the oceans, land, fresh water bodies, and atmosphere. They make it possible for us to observe changes to the planet we live on, from the vantage point of space.

For example, NASA satellites are responsible for providing the first global measurements of aerosols in our atmosphere and for understanding ozone. NASA satellites from the GRACE and ICES missions discovered unexpected rapid changes in Earth's great ice sheets, and the Jason-3, OSTM/Jason-2, and Jason-1 missions have recorded a sea level rise of an average of three inches since 1992.[1]

NOAA and NASA's Earth science programs have provided complementary, unprecedented understanding of how the ocean biology is changing. Europe, China, Japan, South Korea, India, and Argentina have developed and are developing new sensors to follow in NASA's footsteps – and yet at NASA, missions like MODIS and PACE are crucial to design the next generation of ocean sensors.

The core funding for academic research at universities, as well as programs within the various NASA centers, have all played integral roles in our understanding of our unique and rapidly changing world. NOAA's programs play a complementary and critical role as these technologies move into sustained operations through an Integrated Ocean Observing System.

It is imperative that you continue funding Earth science programs at NASA and NOAA.

Preserve Scientific Integrity
There is increasing concern among the scientific community that politics could interfere, stymie, or even silence crucial scientific observations.

We know you share our pride for this country, and a big part of what makes us great is our freedom of speech and our freedom of expression. Politics should never interfere with scientific information being shared with the public.

We are calling on you to foster a strong and open culture of science. We believe in adhering to high standards of scientific integrity and independence. We support legislation, such as that introduced by Florida Senator Bill Nelson, which calls for scientific integrity -- principles for federal agencies including impartiality, freedom from interference, and respect for the scientific process.

Recognize Coastal Properties At Risk
Aptly dubbed, "Our nation's gateway to exploring, discovering, and understanding our universe," NASA's Kennedy Space Center and Cape Canaveral are home to multiple launch pads, several of which are directly on the coast and vulnerable to rising seas, erosion, and storm surge. Please recognize that NASA cannot fulfill its mission to explore outer space if rising seas damage NASA facilities.

Many Florida properties, including Mar-a-Lago - the Winter White House - are vulnerable to sea level rise. If we do nothing to address climate change, we may see

---

[1] NASA Global Climate Change – "Taking a global perspective on Earth's climate: The Earth Observing System era," https://climate.nasa.gov/nasa_science/history/

a foot or more of sea level rise by 2060. America must be vigilant and take action to reduce greenhouse gas emissions.

Climate change can be viewed as a threat or as an opportunity. NOAA and NASA both play a crucial role in helping us to understand those risks. We are confident that the many discoveries accomplished thus far are only the beginning. With continued research, Americans can better understand future challenges and find ways to solve them.

Thank you for your time and consideration.

Sincerely,

Todd Albert, Ph.D.
Chief Executive Officer
Nebular Design, LLC

Senthold Asseng, Professor
Agricultural & Biological Engineering Department
University of Florida

Donald Axelrad, Ph.D., Assistant Professor
Institute of Public Health
Florida A&M University

Ray Bellamy, M.D.
Tallahassee Orthopedic Clinic
Tallahassee, Florida

Leonard Berry, Ph.D.
Emeritus Professor of Geosciences
Florida Atlantic University

Henry Briceño, Ph.D., Research Professor
Southeast Environmental Research Center
Florida International University

Kristen Buck, Assistant Professor
College of Marine Science
University of South Florida

Jeff Chanton, Professor
Department of Earth, Ocean and Atmospheric Science
Florida State University

Eric Chassignet, Professor, and Director
Center for Ocean-Atmospheric Prediction Studies (COAPS)
Florida State University

Allan Clarke, The Adrian E. Gill Professor of Oceanography
Department of Earth, Ocean and Atmospheric Science
Florida State University

James G. Douglass, Ph.D., Assistant Professor
Department of Marine and Ecological Sciences
Florida Gulf Coast University

Marc E. Freeman, Ph.D.
Lloyd Beidler Professor of Neuroscience Emeritus and
Distinguished Research Professor Emeritus
Florida State University

David Hastings, Professor
Marine Science and Chemistry
Eckerd College

Barry Heimlich, Vice Chair
Climate Change Task Force
Broward County

Ben Kirtman, Professor
Department of Atmospheric Science
Rosenstiel School of Marine and Atmospheric Sciences
University of Miami

Dr. Marguerite Koch, Full Professor
Department of Biological Sciences
Charles E. Schmidt College of Science
Florida Atlantic University

David Letson, Professor
Department of Marine Ecosystems and Society
University of Miami

James MacDonald, Associate Professor of Geology
Marine and Ecological Sciences
Florida Gulf Coast University

Heather Mason Jones, Professor of Biology
Department of Biology

University of Tampa

Mason Meers, Ph.D., Professor of Evolutionary Biology & Anatomy
Department of Biology
University of Tampa

Vasu Misra, Associate Professor of Meteorology
Department of Earth, Ocean and Atmospheric Science
Florida State University

Frank Muller-Karger, Professor
Institute for Marine Remote Sensing/IMaRS
College of Marine Science
University of South Florida

John H. Parker, Emeritus Professor of Environmental Science and Chemistry
Department of Earth and Environment
Florida International University

Randall W. Parkinson, Ph.D., P.G.
Research Faculty Affiliate, Institute for Water and Environment
Florida International University

Joelle Richard, Assistant Professor of Marine Sciences
Department of Marine and Ecological Sciences
Florida Gulf Coast University

Ron Saff, M.D.
Allergy and Asthma Diagnostic Treatment Center
Tallahassee, Florida

Michael Savarese, Professor of Marine Science
Department of Marine & Ecological Sciences
Florida Gulf Coast University

Philip Stoddard, Ph.D.
Department of Biological Sciences
Florida International University

John C. Van Leer, Associate Professor
Department of Ocean Sciences
Rosenstiel School of Marine and Atmospheric Sciences
University of Miami

Linda Walters, Pegasus Professor

Department of Biology
University of Central Florida

Harold R. Wanless, Professor and Chair
Department of Geological Sciences, College of Arts and Sciences
University of Miami

John Weishampel, Professor
Department of Biology
University of Central Florida

Disclaimer: *The views and opinions expressed in this letter are those of the individuals and do not necessarily reflect the official policy or position of their respective organizations.*

South Florida Resilient Redesign

City of Delray Beach
Village by the Sea

Concepts

- Water orientation
- Water as amenity
- Maximize water storage
- New structures on piers
- "Complete Streets"
- "Complete Waterways"
- Transportation improvements
    - Beach to mainland
    - FEC passenger opportunity
    - Monorail or trolley
    - Future personal transport
    - Multi-modal system

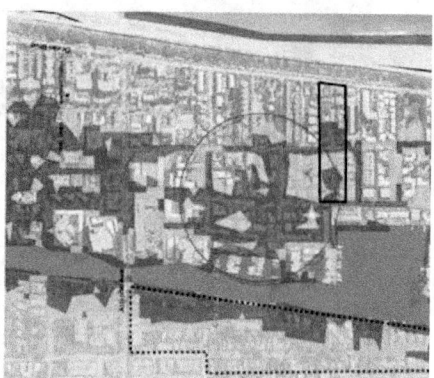

Concepts

- **Nassau district**
    - Impact on western 2/3
    - Emphasis on historic context
    - Raise structures in place
- **Marina district**
    - Poor drainage
    - Persistent flooding and street washout at ICW
    - Convert Marine Way to a bio-swale
    - Raise structures in place
    - Relocate at-risk structures to suitable context for a new historic district

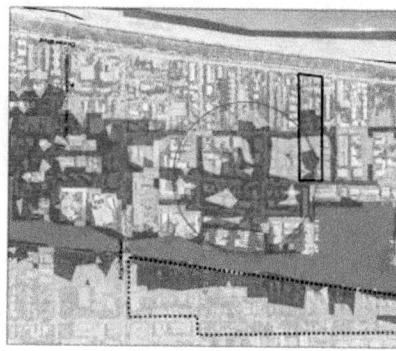

Southeast Florida Resilient Redesign II
An Southeast Florida Regional Climate Change Compact Initiative

Background

The Southeast Florida Resilient Redesign began July 2014, when the Southeast Florida Climate Change Compact (the Compact) collaborated with the Dutch Consulate in Miami to work with local practitioners and stakeholders and experts from the Netherlands to propose resilient design strategies which could serve as models of resilience for communities throughout the south Florida region. The emphasis being on the integration of design solutions into future development and redevelopment projects, and in advance of a major climate disruption. This partnership resulted in *South Florida Resilient Redesign*.

Three characteristic south Florida landscapes were chosen to serve as model sites with the intent to develop transferable design solutions for similar sites across the region, these included: Alton Road in Miami Beach, an area of Sweetwater in Miami Dade and Dania Beach Blvd. in Dania Beach, each representing a barrier island site, suburban site and commercial corridor, respectively.

On the first day, visiting experts and local stakeholders made site visits, learning about the locations on various levels, including details pertaining to cultural, economic, social, historic, topographic, elevation, infrastructure and building stock considerations.

The larger event was held on days two and three and included the participation of nearly 50 professionals with diverse expertise, including water managers, architects, engineers, parks managers, planners, hydrologists, and engineers. In a charrette style format, the group began with a review of the south Florida landscape and development history and then began to consider how to design a resilient community in each of the select settings in response to climate change and natural hazards, while considering area resources or constraints, social dynamics, compatibility with the community's vision and economics, water management infrastructure and implications to neighboring communities.

Other considerations were historic preservation, uniformity, aging infrastructure, evacuation routes, soil permeability, flood hazards, area transit and connections.

The design results were shared with an audience of stakeholders on the final day and then presented at the Southeast Florida Regional Climate Leadership Summit held in October 2014. As a result of this initial collaboration, the City of Dania Beach is now working with Broward County on further refining planning scenarios and design recommendations under an EPA planning grant.

Resilient Redesign II

Resilient Redesign will be held again on July 19-22, 2015. The Compact is pleased to partner this year with the Florida Climate Institute, with the specific engagement of Florida Atlantic University, Florida

International University and the University of Miami. Similar to the first Resilient Redesign, the experts and stakeholders will develop potential design solutions to the evolving urban challenges of climate change and natural hazards.

Teams of experts including architects, landscape architects, engineers, developers, urban designers, as well as experts in hazard preparedness, water management/hydrology/geology, spatial planning, policy, transportation, historic preservation, park management and communications will again be brought together to design solutions.

Sites chosen for this year are locations in Key West in Monroe County, the City of Hollywood in Broward County, and Delray Beach in Palm Beach County. On July 19, the teams will provide an optional local site visit. On July 20 and 21, the teams will come together to undertake the creative process in developing site specific resilient design solutions at the FAU Davie Campus. The presentations will be held on the morning of July 22 at the FAU Davie Campus.

Site Briefings

Hollywood, Hollywood Blvd

The Hollywood Boulevard site within the City of Hollywood, Broward County, Florida is representative of urban development in South Florida. As in other coastal urban areas, waterfront property and transportation infrastructure are subject to, or at risk, of flooding as a result of rain events, extreme high tide events, storm surge and sea level rise. The site's flood protection and drainage infrastructure is in need of retrofitting and overall the community is in need of a resilience and adaptation strategy that will sustain the waterfront community with limited resources.

The community surrounding the east end of Hollywood Boulevard includes the edge of a central business district in active redevelopment, waterfront residential property and commercial property on the barrier island. The selected Hollywood Boulevard site extends 1.7 miles east to west and 1 mile to the north and south. The majority of the site is very low-lying and below an elevation of 4 feet NAVD. The western edge of the site borders the elevated coastal ridge (an ancient sand dune) which runs north to south at an elevation of approximately 10-11 feet NAVD. The site overlaps the city's two Community Redevelopment Agency (CRA) districts, Downtown Hollywood and Hollywood Beach. As such, there are contrasting socioeconomic conditions between the blighted areas in transition and the high value waterfront residential properties in the center. However, overall, the city intends to redevelop the area into a "world class coastal destination" and "classic American downtown with an international accent."

The residential area includes Hollywood Lakes which is the oldest neighborhood in the city, dating back to the 1920's. Several iterations of redevelopment have occurred in the community, initially when recovering from devastation from the 1926 Hurricane, Hurricane Cleo in 1964, Hurricane Betsy in 1965, Hurricane Andrew in 1992, and in response to high demand for coastal property (2000s). In many cases, property redevelopment did not necessarily result in the reconstruction of seawalls or placement of additional fill material to raise the property, leading to the current conditions of temporary flooding along waterfront boundaries in response to high tides and elevated groundwater levels. In order to counteract flooding, the city and its residents are implementing temporary site-specific solutions.

The commercial property along the barrier island faces pressure both from beach erosion and surge along the coastline and tidal flooding along the intracoastal waterway. In general, infrastructure along the commercial corridor is aging and ready for redevelopment. Existing development lies seaward of the Coastal Construction Control Line leaving it vulnerable to storm-induced damage and erosion. Furthermore, the area has limited protection by natural infrastructure since the majority of the coastline within the site area lacks vegetated dunes in favor of direct ocean views and direct access to the beach from the street. Redevelopment could provide an opportunity to increase resilience if existing infrastructure and development is redesigned.

Key West, Salt Ponds Area

The City of Key West sits an average of 4.7' (is it 4.7 feet or 4 feet 7 inches, not the same thing) above sea level. Home to ~24,600 people, it entertains over 3 million visitors a year, and enjoys international acclaim as a tourist destination. Because of our topography, geography and geology, the City of Key West has a long history of hazard mitigation and adaptation.

Infrastructure: Situated in the middle of the Florida Keys National Marine Sanctuary, protecting water quality is very important for our Keys-wide economy. In 1989, Key West was the first in the Keys to incorporate central sewer, and did so via pump technologies because gravity fed systems were not appropriate. Via storm drains, the island has 11 intersections that inundate monthly during full moon high tides, and are easily exacerbated with less than one inch of rain. The City's 2012 Stormwater Master Plan utilized LIDAR data to better address our flooding issues, and committed $5M of the next few years of Capital Improvement Funding for the installation of pump assist injection wells, elimination and/or retrofit of outfalls, installation of injection wells, and retrofit of existing injection wells.

Policies: The 2012 Comprehensive Plan included 56 new items of sustainability, including adaptation language specifically referencing Adaptation Action Areas, raising of buildings, green infrastructure, natural areas resiliency and other innovative concepts. In the Land Development Regulations (LDRs), the City's recently adopted Building Permit Allocation System only gives new residential unit permits to homes that build 1.5' above flood, obtain green certification and install 1,000-gallon cisterns. The City has recently procured a consultant to do an overhaul of our LDRs, with half of the funds coming from a Sea Grant grant specifically for adaptation policy development. Mitigation funding mechanisms for transportation, tree canopy, stormwater, etc are specifically desired, as is higher allowable densities in order to make affordable housing developments more feasible.

The study area is comprised of <600 acres of land and water in the low-lying southeast corner of the island of Key West. The area is bounded to the north by Flagler Boulevard, and the Riviera Canal, to the east and south by the Atlantic Ocean, and to the West by White Street. Natural areas consist of the Salt Ponds, Indigenous Park, Berg & Kitsos tracts, the Little Hamaca Park complex, and .6 miles of beach.

There are three main components we wish to explore through the Resilient Redesign process: a low lying suburban neighborhood, the County airport and the eroding beach / state road that protects it all.

1) The 32 block "Indigenous Park neighborhood" holds the City's heaviest concentration of greatest flood damage, costing FEMA $127M in flood claims since 1983. Ranging in elevation from AE6 to AE10, the single family homes are primarily on-grade, concrete block, single story structures, with >90% built before 1970. The area experiences significant flooding with average rain events, especially during high tides. While the City has recently revised our LDR's to incentivize home raising, these homes cost significantly higher to raise ($100,000 versus $20,000 for a wood frame home). There is interest in piloting a modified floodproofing program here.

2) The Monroe County International Airport was built in the middle of a wetland. Hosting four major airlines, it serves 362,000 passengers annually. It was the most affected airport in the Compacts four County vulnerability study, with 50% of its property affected with only one foot of sea level rise and 94% affected by three feet of SLR. Inundation isn't' the biggest problem, however. Long before water rises over the tarmac, the saturated base material of the runway can compromise the strength and integrity of the landing strip.

3) The beaches in Key West are small and well-loved. Rest Beach and Smathers Beach, sandwich the Berg and Kitsos natural areas, along with a few condos, stretching for a total of .6 miles of sand. While Smathers gets replenished from FEMA after large storms on a roughly 5 year average and gets as much as 150 feet at its widest, Rest Beach has not been replenished for 20 years and in many cases is undermining boardwalks, trees and structures. The remaining 1.8 miles of shoreline consists of the riprap used to protect South Roosevelt Boulevard, the southernmost stretch of A1A. An important evacuation route for the island, South Roosevelt is buffeted by mangroves on the interior, but is mostly unprotected seawall on the Atlantic side.

Delray Beach, Intracoastal Basin

The Delray Beach Resilient Redesign Study Area (study area) encompasses the eastern portion of the Town of Linton (platted in 1896) and a portion the Coastal Planning Area established in the Coastal Element of the Local Government Comprehensive Plan. The study area represents the essence of what is Delray Beach; the Village by the Sea. The long term sustainability and resilience of this area is essential to managing growth and preserving the charm.

The study area is bounded on the north by George Bush Boulevard, south by 10th Street, west by North Federal Highway (6th Avenue), and east by the Erosion Control Line to the east of Highway A1A (Ocean Boulevard). The eastern (barrier island) and western portions of the study area are split by the Atlantic Intracoastal Waterway. The 703 acre study area is comprised of 613 acres of land and 96 acres of water. As in other coastal urban areas, waterfront property and transportation infrastructure are subject to, or at risk of, flooding as a result of rain events, extreme high tide events, storm surge and sea level rise. The site's flood protection and drainage infrastructure is in need of retrofitting and overall the community is in need of a resilience and adaptation strategy that will sustain the waterfront community with limited resources.

The land in the study area is essentially built-out and the land use is overwhelmingly residential with more than half the area zoned for single family detached housing. Commercial land uses are found along Atlantic Avenue (Central Business District), Ocean Boulevard, and George Bush Boulevard. Infrastructure improvements and redevelopment within the study has been an ongoing process since the beginnings of the early settlement. The existing public infrastructure includes water, sewer, and reclaimed water systems, storm drains, and streets to serve the built out character of the area.

The economic character of the study area is generally residential, with limited commercial development, providing support for both residents and tourists. The economy is reflective of the economy of the City as a whole, and in turn, of the region. It is a service economy based upon full-time residents. There is, however, a significant tourist and seasonal component within the economy that is oriented toward the beach resource and downtown.

There are two locally designated Historic Districts in the Study Area. They are the Nassau Street Historic District and the Marina Historic District. There are ten individually designated structures on the local historic register. The City has a Historic Preservation Ordinance which is administered by the Historic Preservation Board. That ordinance requires issuance of a Certificate of Appropriateness by the Board prior to modification or new construction on properties in Historic Districts and on individually designated sites.

Natural resources consist of the beach/dune ecosystem and the offshore reef. In Delray Beach, there is a single offshore coral reef, in approximately 55 feet of water. Long term monitoring indicates this reef to be one of the healthiest and most diverse reef environments in the region. Between the reef and shore is sand bottom, there are no hard bottom areas or reef or rock outcrops.

The City has one of the premier beach erosion control and nourishment programs in the State of Florida. In 1973, the City constructed an initial beach restoration, placing sand by hydraulic dredging from an offshore borrow area. Over the years maintenance and storm mitigation nourishment projects have been constructed. The beach nourishment project has successfully provided storm protection for upland property. Since 1973, there has been no damage to upland property due to erosion or storm damage.

Beach nourishment has proven to be a very satisfactory solution to long term erosion by providing a flexible buffer to the impact of storm waves. In addition to recreation and storm protection, beach nourishment has recreated a habitat for nesting sea turtles.

The City has recreated a dune system at the Municipal Beach through a long term program of reconstruction and maintenance. Over the years, this has resulted in a distinct fore dune and primary dune, vegetated in native species. Adjacent to privately owned properties, remnants of the natural dune have also survived. Many property owners have undertaken vegetation projects similar to those of the City to recreate a vegetated fore dune.

Technically, there is no estuarine environment in the Delray Beach Resilient Redesign Study Area. The Atlantic Intracoastal Waterway has been channelized throughout the City, and most of the shoreline is protected by seawalls. The natural areas which remain are not estuarine in a technical classification. However, the City considers the Atlantic Intracoastal Waterway to be more than a transportation route and its estuarine features (living shorelines) should be restored and protected.

Senator NELSON. Thank you, Dr. Berry.
Mr. Hedde.

STATEMENT OF CARL G. HEDDE, CPCU, HEAD, RISK ACCUMULATION DEPARTMENT, MUNICH REINSURANCE AMERICA INC.

Mr. HEDDE. Good afternoon. Thank you for inviting me to testify.

As Senator Nelson mentioned, I work for a company, Munich Re, that allows us to talk about climate change.

Today, I want to share some thoughts on the potential financial impact posed by climate change and on resiliency steps that society must undertake to mitigate the human and financial toll of our changing climate.

Let me begin with a quote from Dr. Peter Hoppe, Head of the Munich Re Corporate Climate Center, and a leading climate change expert. He said, "A look at the weather-related catastrophes of 2016 shows the potential effects of unchecked climate change. Of course, individual events themselves can never be attributed directly to climate change. But there are now many indications that certain events, such as persistent weather systems or storms bringing torrential rain and hail, are more likely to occur in certain regions as a result of climate change."

The insurance industry relies heavily on historical loss information to make business decisions. However, the use of historical loss data assumes that the risk we see today is the same that it was in the past. This is not always the case. As an industry, we also rely heavily on weather and climate data produced by NOAA.

Losses from weather catastrophes in the U.S. have increased in both frequency and severity over the past 4 decades. In the U.S., socioeconomic changes have played a substantial role in this increase, but do not explain the entirety of the changes. It is likely that the changes in climate, whether from natural variability or due to man's influence, are playing a role in these trends.

During 2016, the U.S. experienced approximately 91 large natural catastrophe events. Of the 91 events, only two were caused by earthquakes. The other 89 events have a climate connection.

The economic loss for these 91 events totaled approximately $43.9 billion U.S. dollars, of which $23.8 billion, or 55 percent, was covered by insurance. The difference is the amount that is not insured and thus directly borne by U.S. citizens in the form of retained loss, cost of disaster relief, or lost economic opportunity.

Within the U.S., approximately 50 percent of all losses caused by large natural catastrophe losses are historically not covered by insurance. In the U.S., related events such as hurricanes and other wind-related events have a higher penetration of insurance coverage as compared to earthquake and flood events.

Human safety is our greatest concern when natural cat events occur. During 2016, 260 lives were lost due to severe weather events in the U.S. Just weeks ago on March 30, a mother and her 3-year-old daughter lost their lives in a relatively weak EF–1 tornado with estimated winds of about 100 miles per hour. The tornado occurred in Breaux Bridge, Louisiana, placing their mobile home in the HUD Zone 2, with mobile home design speed requirements of around 100 miles per hour. Unfortunately, it appears that this mobile home was likely built before HUD standards were adopted, and it appears to have been placed on cinder blocks without tiedowns.

Munich Re feels that if mobile homes were required to be properly sited and designed to the stronger HUD Zone 3 standards, this tragic loss of life could have been avoided. So we recommend that all mobile homes in the U.S. be designed and installed at the Zone 3 standards, and those would be built to withstand about 110 miles per hour.

These tragedies can be significantly reduced or even avoided if proper building codes and enforcement are in place. The citizens of Florida enjoy some of the most stringent building codes in the U.S.

After Hurricane Andrew in 1992, Florida officials strengthened both the building codes and building code enforcement. It is time for building codes to be strengthened across the U.S.

As a Nation, we must build stronger homes. Most of our homes and businesses are not built to a changing climate and the weather that a changing climate might bring. The cost in terms of lost lives, and damage to homes and businesses, and community post-event viability is devastating.

So in summary, we must address climate change on multiple fronts. Munich Re supports a smart, balanced approach that protects the public but does not stifle business or innovation. Researching and addressing the genesis of climate change is one step. Preparing our Nation for the impacts of a changing climate must happen concurrently.

National preparation must keep citizens safe in the face of the natural disasters we are experiencing. Safety starts in our homes, and extends to our schools and businesses. It is in the mutual interests of the Federal Government and the insurance industry to partner to find solutions in the areas of adaptation and risk.

Thank you again for allowing me to testify today.

[The prepared statement of Mr. Hedde follows:]

PREPARED STATEMENT OF CARL G. HEDDE, CPCU, HEAD, RISK ACCUMULATION DEPARTMENT, MUNICH REINSURANCE AMERICA INC.

Good afternoon and thank you for inviting me to testify. I am Carl Hedde, Head of the Risk Accumulation Department at Munich Reinsurance America Inc., one of the largest reinsurers in the United States.

Founded in 1917, we have over 1,000 employees serving our clients in the U.S. Our parent company, Munich Re, is one of the world's leading reinsurers.

Today I will share thoughts on the potential financial impacts posed by climate change, and on resiliency steps that society must undertake to mitigate the human and financial toll of our changing climate.

Let me begin with a quote from Dr. Peter Hoppe, Head of the Munich Re Corporate Climate Center, and a leading climate change expert. Dr. Hoppe said, "A look at the weather-related catastrophes of 2016 shows the potential effects of unchecked climate change. Of course, individual events themselves can never be attributed directly to climate change. But there are now many indications that certain events—such as persistent weather systems or storms bringing torrential rain and hail—are more likely to occur in certain regions as a result of climate change".

The insurance industry relies heavily on historical loss information to make business decisions. However, the use of historical data assumes that the risk we see today is the same as it was in the past. This is not always the case. As an industry, we also rely on weather and climate data produced by NOAA.

Losses from weather catastrophes in the U.S. have increased in both frequency and severity over the past four decades. In the U.S., socioeconomic changes have played a substantial role in this increase, but do not explain the entirety of the changes. It is likely that changes in climate, whether from natural variability or due to man's influence, are playing a role in these trends.

During 2016, the U.S. experienced approximately 91 large natural catastrophe—or "Nat cat" events, after removing small scale events. Of the 91 events, only two were caused by large earthquakes. The other 89 events have a climate connection. The frequency of earthquakes is remaining relatively constant.

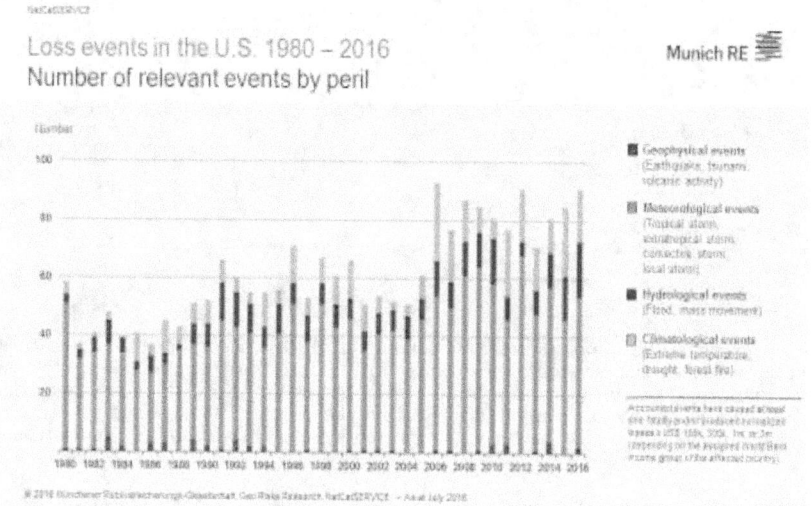

The upward trend in regard to losses from weather catastrophes is also apparent in the economic and insured losses for some weather-related perils, such as severe tornado and hail outbreaks in the U.S.

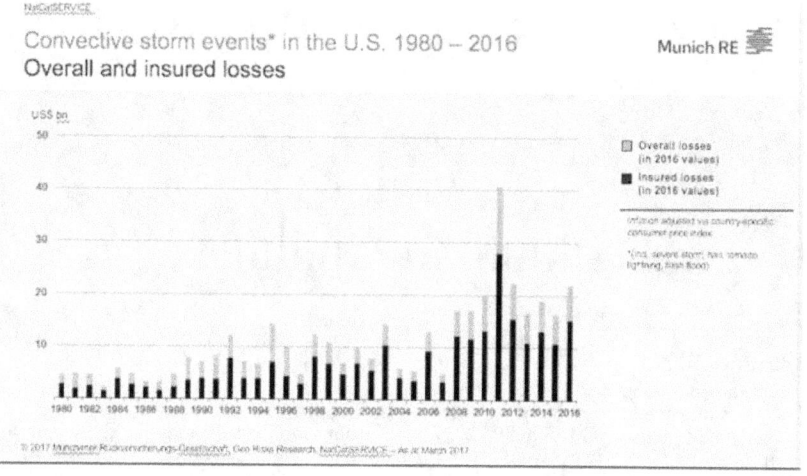

The economic loss total for these 91 events totaled approximately $43.9 billion U.S. dollars, of which $23.8 billion, or 55 percent was covered by insurance. The difference between Economic and Insured Losses is the amount that is not insured and thus directly bourn by U.S. citizens in the form of retained loss, cost of disaster relief or lost economic opportunity.

As of January 4, 2017	Number of Events	Fatalities	Estimated Overall Losses (US $m)	Estimated Insured Losses (US $m)*
Severe Thunderstorm	43	40	19,000	14,000
Winter Storms & Cold Waves	7	55	1,700	1,000
Flood, Flash Flood	19	83	15,000	4,300
Earthquake & Geophysical	2	-	Minor losses	Minor losses
Tropical Cyclone	2	52	7,000	3,500
Wildfire, Heat Waves, & Drought (ongoing drought condition without loss estimation for the half year)	18	32	1,200	1,000
Totals	91	262	43,900	23,800

Within the U.S., approximately 50 percent of all losses caused by large natural catastrophes are not covered by insurance. In the U.S., weather related events such as Hurricanes and other wind related events generally have a higher penetration of insurance coverage as compared to Earthquake events. The coverage penetration for flood protection is also very low, and presents a critical exposure for our citizens.

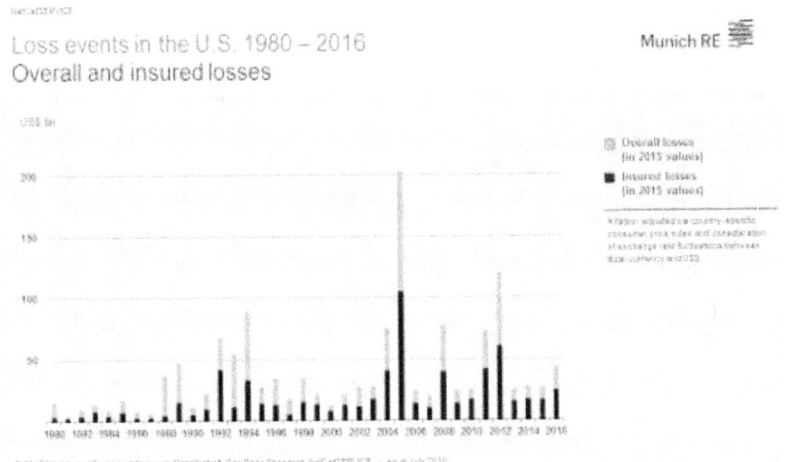

Adaptation

Human safety is our greatest concern when natural cat events occur. 262 lives were lost due to severe weather events in the U.S. in 2016.

Just weeks ago, on March 30, 2017, a mother and her 3 year old daughter lost their lives in a relatively weak EF–1 Tornado with estimated winds of 100 MPH. The tornado occurred in Breaux Bridge, LA, placing their Mobile Home in the HUD Zone 2, with Mobile Home design speed requirements of 100 mph. Unfortunately it appears that this particular mobile home was likely built before HUD standards were adopted, and appears to have been placed on cinder blocks without tie-downs. Munich Re feels that if the Mobile Home was required to be properly sited and designed to the stronger HUD Zone 3 standards, this tragic loss of life could have been avoided. Munich Re recommends that all Mobile Homes in the U.S. be designed to the stronger HUD Zone 3 standards, which are built to withstand winds of 110 MPH.

We constantly hear tragic stories of communities being devastated by severe weather events. These tragedies can be significantly reduced or even avoided if proper building codes and enforcement are in place. The citizens of Florida enjoy some of the most stringent building codes in the U.S. After Hurricane Andrew in 1992, Florida officials strengthened both the building codes and building code enforcement. It is time for building codes to be strengthened across the U.S. . . .

Taking Action

As a nation, we must build stronger homes. Most of our homes and businesses are not built to withstand the weather we have today, let alone the weather a changing climate might bring. The cost—in terms of lost lives, damage to homes and businesses, and community post-event viability—is devastating.

Munich Re supports a non-profit organization called "IBHS"—the Insurance Institute for Business and Home Safety. The insurance industry and IBHS have been conducting research into, and promoting stronger building codes and construction practices. Much of the findings are incorporated into the IBHS Fortified Program which essentially promotes Code+ construction standards.

Munich Re is committed towards the goal of making our country's building stock more resilient than it is today. For example, Munich Re funded the development of a "tablet app" in partnership with the IBHS that can help users build or retrofit a home to IBHS Fortified standards. The "app" is available for free on the iTunes app store, and we encourage all attendees at today's meeting to test it out and distribute it to your constituents.

In summary, we must address climate change on multiple fronts. Munich Re supports a smart, balanced approach that protects the public, but does not stifle business or innovation. Researching and addressing the genesis of climate change is one step. Preparing our Nation for the impacts of a changing climate must happen concurrently. National preparation must keep citizens safe in the face of the natural disasters we are experiencing. Safety starts at our homes, and extends to our schools and businesses. It is in the mutual interest of the Federal Government and the insurance industry to partner to find solutions in the areas of adaptation and risk transfer.

This makes absolute sense from a macroeconomic perspective, as lower subsequent losses will generate savings of several times the investment. Most importantly—these solutions can protect human lives.

Thank you again for providing this opportunity for me to testify.

Senator NELSON. Thank you, Mr. Hedde.
Dr. Jurado.

STATEMENT OF DR. JENNIFER L. JURADO, CHIEF RESILIENCE OFFICER AND DIRECTOR, ENVIRONMENTAL PLANNING AND COMMUNITY RESILIENCE DIVISION, BROWARD COUNTY, FLORIDA

Dr. JURADO. Thank you. Good afternoon, Senator Nelson, Congressman Deutch, and Mayors. It is a great honor to speak with you regarding the very real challenges facing our county and Southeast Florida as a whole regarding climate change.

Our region is working aggressively to address the reality and the impact that sea-level rise is affecting our coastal counties. Not only is the global science clear, but regional data and local impacts align unambiguously with the predictions.

We know that the sea level has risen and continues to rise at an ever-accelerating rate. Like many regions in the states, we are already grappling with the impacts.

However, in South Florida, our circumstance is unique given the expansive impacts on our region's drainage, our flood control, and our water supplies.

These are not examples of future risk, but are reality today. Tidal flooding is not just a mere nuisance, but an agonizing recurrence for many of our communities.

Seasonal flooding isolates neighborhoods, restricts commerce, and disrupts essential services.

As you see in the many photos, storm water systems and streets fill with seawater as walls are overtopped. Coastal waters backflow through pipes and storm drains. Marinas now funnel water into streets and nearby neighborhoods. And businesses, as Congressman Deutch has seen, must seal their front doors with caulk in order keep the water at bay.

These are the visible impacts, but the region's vulnerabilities extend well beyond the coastal corridor. Rising seas will contaminate 40 percent of the coastal wellfield capacity in Broward County alone. Our regional flood management system, the Central and South Florida Flood Control Project, is losing capacity as rising seas restrict discharges at tidal structures, preventing flood relief during rainfall events. Groundwater elevations are rising, reducing soil storage and compounding flooding miles inland.

The situation is challenging but it has also marshaled incredible leadership across our region. As you have heard, in 2010, the Southeast Florida Climate Change Compact was formed by Broward, Palm Beach, Miami-Dade, and Monroe counties, inclusive of 108 cities in our region. Through the efforts of the Compact, we have accelerated climate resiliency policy, planning, projects, and aided the realization that immediate action is required.

We understand science. We have developed the tools. Collectively, we are all utilizing the same sea-level rise projection to inform our planning, our consulting services. It is time to invest in infrastructure to reduce risk, improve our communities, and stimulate our economies. Across the region, we have adopted the sea-level rise projection. It is integrated across our processes.

We have pursued strong partnerships with our Federal agencies that have continued to inform the scope, depth, and power of our work, a decade-long collaboration with the USGS to form the hydrologic models. We are utilizing tools developed by FEMA. We are partnering with the Army Corps of Engineers. We work with the EPA on resiliency studies. And we are utilizing a NOAA-funded project to further address resiliency.

These are the investments that are now informing the design standards for drainage systems, higher finished floor elevations, and resiliency for interconnected systems.

In Broward County, we are amending our code of ordinances to formally adopt a future conditions map for the design of all infrastructure in our county. We recognize local communities must shoulder certain conditions and responsibilities. However, Federal and State action are equally important.

Given our location between two national priorities, the Everglades and the Atlantic Coast, our flood protection relies upon upgrades to the Central & South Florida Flood Control Project and the Intracoastal Waterway.

We urge the Federal leadership to prioritize resiliency and infrastructure improvements of this infrastructure. We know the price tag will be large, but these are investments that will further the economies where we already account for one-third of the State GDP, but coastal communities also make up 50 percent of our national economy.

Investment in coastal resilience will bolster economic activity through vast public works projects, through the preservation of economies long term. And, ultimately, this is just the beginning in terms of where these impacts are most visible today. All of our Nation ultimately relies upon the revenues and economies that our coastal communities support.

These impacts are already in motion. We hope that Federal leadership will continue to provide resources to support the science that aids our communities, clean energy that will help harness the magnitude of future changes, and investments in transportation programs where so much of our emissions are generated.

We want to do our share of the national and international work to cut emissions and are eager to partner with the Federal Government in this effort. We thank you for the opportunity and the efforts to bring policy and resources to bear in addressing climate risk and impacts to our community and Nation. Thank you.

[The prepared statement of Dr. Jurado follows:]

PREPARED STATEMENT OF DR. JENNIFER L. JURADO, CHIEF RESILIENCE OFFICER AND DIRECTOR, ENVIRONMENTAL PLANNING AND COMMUNITY RESILIENCE DIVISION, BROWARD COUNTY, FLORIDA

Good morning, Senators, Representatives, and distinguished guests.

I am Dr. Jennifer Jurado, Chief Resilience Officer and Director of the Environmental Planning and Community Resilience Division for Broward County. Thank you for convening this hearing here in Southeast Florida and for bringing attention to the vitally important issues of climate science, coastal risk, and the urgency of action. It is a great honor to speak with you regarding the very real challenges facing Broward County and southeast Florida as a whole. Our region is working aggressively to address the increasingly unavoidable realities of climate change, and the impacts of sea level rise in particular on coastal counties.

Not only is the global science clear, but regional data and local impacts align unambiguously with the predictions.

We know that the sea level has risen and continues to rise at an ever accelerating rate. Like many regions across the United States, we are already grappling with impacts. However, the south Florida circumstance is unique given the expansive impacts of sea level rise on our region's drainage, flood control, and water supplies.

These are not examples of future risk and exposure, but rather the reality we experience in our communities from sea level rise today. Tidal flooding is no longer a mere nuisance, but an agonizing recurrence for many south Florida communities, with water levels that increasingly exceed predictions and place people, businesses, and infrastructure at increased risk.

Seasonal tidal flooding isolates neighborhoods, restricts commerce, and disrupts essential services. Stormwater systems and streets fill with water as seawalls are overtopped. Coastal waters backflow through storm drain pipes, causing water to rush out of storm drains. Marina boat ramps turn into funnels, pouring the ocean onto nearby streets and inundating neighborhoods. Businesses must seal their front doors with caulk to keep the water at bay. Employees and customers use alley entries, donning galoshes or rolling up their pant legs. Many have become practiced at making a mad dash for their vehicles before the parking lot and adjacent roadways become inundated with saltwater for 2–3 hour periods. This condition repeats 12 hours later, and daily over the next week, and again the following month, and so on.

These are the visible effects, but the region's vulnerabilities to the impacts of sea level rise extend well beyond our coastal corridors. Our regional flood management system, the Central & South Florida Flood Control Project (CSF Project for short), is losing capacity as rising seas restrict discharges at tidal structures, compromising flood protection for inland communities miles away from the coast. The South Florida Water Management District has estimated that 18 of these structures are already within six inches of their design capacity. At the same time, due to our porous geology, groundwater elevations are rising in direct response to the rising sea, reducing the ability of the soil to store water and thereby exacerbating flooding.

The situation is immensely challenging, but it has also marshalled leadership from across our region, in an organized effort focused on both climate mitigation and adaptation strategies. In 2010, this regional collaboration was formalized as the Southeast Florida Regional Climate Change Compact, organized by Broward, Palm Beach, Miami-Dade and Monroe Counties, and inclusive of 108 municipalities. The efforts of the Compact have accelerated climate resiliency policy, planning, and projects throughout the region in an unprecedented fashion that has become a model for regions across the globe. There is an urgency in our efforts, and a growing realization across all sectors that the time for preemptive action has already passed and that immediate action is required. We understand the science. We have developed the tools. It's time to invest in infrastructure and updated design standards that will reduce risk, improve our communities, and stimulate our economies.

Our communities are taking action and ownership. Across the region we have formally adopted a unified sea level rise projection as the basis for adaptation planning and integrated that projection into capital, land use, water supply, transportation, emergency management, and capital investment plans. We have pursued strong partnerships with our Federal agencies, relationships that continue to expand in their scope, depth, and power to inform.

This has included:

- A decade-long collaboration with the United States Geological Service in the development of advanced hydrologic models to help assess the impacts of sea level rise on water supplies and flood elevations.
- A resiliency study under the Planning Assistance for States Programs with the United States Army Corps of Engineers to evaluate the future flood conditions, risk reduction measures, and resiliency standards for sea walls.
- A study that will build upon modeling tools developed by FEMA. A collaborative resiliency study with the EPA focused on a coastal municipality.
- A partnership with Deltares, a Dutch water institute, in a NOAA-funded project focused on future flood risk, sectoral interactions, and decision-support for resiliency planning.

The technical investments and projects are now being translated to practical applications as we prepare to update our regulations to address future flood conditions, including revised design standards for drainage systems, higher finished floor elevations, and consistent standards for interconnected and interdependent systems. These investments will help reduce flood risk, including future FEMA losses, and will create a foundation for a more resilient community and economy.

We recognize our local communities must shoulder certain responsibilities to protect our residents, businesses, infrastructure, properties, and environmental resources, and we are aligned regionally in this determination. However, the effectiveness of our regional planning, of all of our investments, is dependent on strong Federal and state partnerships and joint action. Given our location between two vitally important national priorities, the Everglades and the Atlantic coast, our flood protection relies upon federally-supported upgrades to the Central & South Florida Flood Control Project and the Intracoastal Waterway.

So, as we continue to make significant advancements regionally and locally, we urge our Federal leadership to prioritize the building of resilient communities and infrastructure and specifically request that our Federal and state partners undertake a detailed evaluation of the CSF Project and Intracoastal Waterway for flood protection service levels under evolving climate conditions, especially sea level rise, and to develop a comprehensive infrastructure and funding plan to execute the necessary improvements for our region's flood protection.

I mentioned the partnership agreement with the Army Corps to develop recommendations for resilient sea walls. With more than nearly 400 miles of armored and armored shoreline, preliminary estimates suggest a cost between $1 billion and $4 billion. Needless to say, this level of investment is beyond the means of Broward County government alone to finance. But this investment and many others will need to be undertaken, because the no-action alternative will be far more costly.

Of course, seawalls are only part of the solution. Coastal flood control structures—part of the CSF Project—will need to be converted from gravity-based to pump-operated systems. Improvement of each structure is likely to cost around $50 million. Stormwater systems will need to be upgraded. Miami-Beach has advanced a $400 million stormwater improvement plan, just to address sea level rise. Roads will need to be raised and drainage systems will need to be reengineered.

Coastal communities account for nearly 50 percent of our national GDP. Southeast Florida accounts for one-third of our state GDP. Investment in the resilience of these communities is essential to preserving not only local economies, but our national economy. Investments in resilience will provide shared benefits. Deferral is an option we cannot afford.

I would also like to take a moment to acknowledge that we also have an obligation locally, as a state, and as a nation, to reduce greenhouse gas emissions—and this need is just as urgent. Many climate impacts will be unavoidable, given our emissions to date. But our choices today and in the next few decades about energy, transportation, land use, food, and environmental protection will determine if future generations face manageable climate impacts or severely compromised conditions.

In Southeast Florida, we are well aware of our contributions to the problem of climate change. We have some programs in place to reduce our energy and fuel use, but our efforts are admittedly too modest. We face significant challenges—often beyond our direct control—in deploying solar energy, reducing building energy use, and transitioning to a transportation system with more choices and fewer emissions. Fortunately, the Federal Government has been a leader and partner over the last decade, helping local governments and regions to begin to overcome these barriers by offering significant technical assistance, backed by national-level policy decisions which provide great benefit to our vulnerable region. We hope that our Federal leadership continues to provide resources to Federal agencies for climate science, clean energy and transportation programs, and local assistance. We want to do our share of the national and international work to cut emissions, and we will, but we continue to need the Federal Government's help.

Thank you again for this opportunity and for your efforts to bring policy and resources to bear in addressing the risk and impacts of climate change on our communities and nation.

Senator NELSON. Thank you.

I want to commend all of you. I have never seen witnesses pay attention to the 5-minute rule.

[Laughter.]

Senator NELSON. So thank you.

By the way, if anybody is wondering why we do not have somebody that has the opposite opinion here, we have had that plenty of times in front of this Committee. The case today is the Full Committee. But a year ago, we had the Science and Space Subcommittee chaired by Senator Cruz of Texas, and the minority was allowed one witness. There were four witnesses by the majority, and all of them were denying that the climate was changed.

We selected as the minority witness the former head of meteorology for the United States Navy, a retired Admiral. And of course, he presented in livid and living detail a lot of the points that you all have pointed out.

So there has been a fair hearing of all parts of the issue saying that it is fair and balanced. However, when you talk about scientific attitudes, it is another thing. So, thank you, you are particularly good on this.

Now, what is the procedure here? The Committee will ask some questions of the witnesses. I am going to defer to our witnesses up here, and I will do some cleanup.

So let me turn to Congressman Deutch.

Mr. DEUTCH. Thanks very much, Senator.

Thanks so much to the witnesses. I think the answer to why are there not other views, Dr. Kirtman, you provided the answer, which I will refer to over and over again, and that is "multiple lines of evidence," and it is appropriate for us to be hearing from all of you who have seen and understand and appreciate those multiple lines of evidence.

I want to ask a question to everyone on the panel, and it has to do with the President. And I would like you, if the President were sitting here across from you, instead of as close as he is on a regular basis, as our Mayors know, here in Palm Beach County—I do not want to focus on the concerns that a lot of us have about statements from some in the Administration about climate change and how they view climate change. I want to focus on the President's campaign promise and repeated statements since being elected that he wanted to make a massive investment in infrastructure.

And I want to take advantage of having all of you here, because you all, in one way or another, have talked about the importance of resiliency and adaptability and preservation and mitigation.

So if the President asked you what kinds of investments in our big infrastructure plan that we all hope is coming, what kinds of investments can we make that will sustain our economy, generate economic growth, and, at the same time, help us address the very real issues that we are grappling with that have been the focus of this hearing, what kinds of investments would you tell him to make?

Dr. Jurado, why don't we start with you?

Dr. JURADO. Sure, I would be glad to address that question.

First of all, beginning with the Central and South Florida flood control system, we know that 18 of the salinity control structures on the system are within 6 inches of their design capacity. Those gravity control structures need to be retrofit with pumps, and we need a comprehensive plan on how we will maintain flood elevations for flood protection for an entire community through modified operation and investment in that system.

During the 2015 high tide events, the Intracoastal Waterway has been overtopped in Broward County and Hollywood. Last year, the high tide events were 1 inch beneath the top wall of the Intracoastal Waterway. This is Federal infrastructure well beyond the capacity of any community to be able to upgrade. We are working with the Army Corps of Engineers on recommended top elevations for our seawalls under future climate conditions. We should use that information to inform investments on those barriers to coastal flooding.

Seawalls in Broward County, we have 350 miles of shoreline with all the finger canals and waterways in the county. Those will collectively have to be upgraded to a common top elevation. Easily, that price tag is $1 billion to $4 billion, depending on the structural integrity of those seawalls.

And you have all the municipal storm systems and roadways that can potentially be addressed at the same time.

Those are the key pieces of infrastructure that I suggest we might begin with.

Mr. DEUTCH. Thank you very much.

Mr. Hedde?

Mr. HEDDE. Sure. As a global insurer, reinsurer, we wind up paying losses across a wide spectrum of properties, whether they are residential or commercial. And we are strong supporters of stronger and large investment in infrastructure. We focus on the infrastructure that ultimately will protect these businesses and homes that we are working in.

On a micro level, listening to us, we talk a lot about protecting homes, better building codes. We are a member of the IBHS, the

Insurance Institute for Business & Home Safety, focusing on keeping people safe with broader, better protection across the economy.

Mr. DEUTCH. Dr. Berry?

Dr. BERRY. Yes, I would focus on the kind of infrastructure. And just like after hurricanes, we have a tendency to put back what was there before. I would like to see any new infrastructure from roads to waterways and so on built with an appreciation of climate change.

Let's have roadways that cannot move water off them but can add water into them. Let's build water and roadways that are compatible with one another.

What we often do is put in a highway, and then we have to add drains and we have to add all kinds of other things to get rid of the water up there.

Let's look at an integrated way to use a big $1 trillion infrastructure bill to create a new moon landing kind of approach to it. Let's not do what we have done. Let's create a new image of what is the infrastructure of the future, and build our electronic infrastructure into this too. We always think of roads and rails. What about the electronic infrastructure that we need? How should that be? How do we look at that 20 years ahead?

So a futuristic view is what I would appreciate.

Mr. DEUTCH. Thank you.

Dr. Kirtman, any thoughts?

Dr. KIRTMAN. It is difficult for me to expand on what my colleagues have said. That is a wonderful list. I hope that all happens.

The one thing is I would like to underscore some comments that Len said. I think it is critically important that we take a holistic view when we think about infrastructure, and that holistic view has to include multiple time horizons and multiple inputs.

So we really need to be very, very careful that we implement a no-regrets strategy, that we do not do anything that we regret 10 years or 20 years or 30 years from now, and we also need to be very careful that we take in all the best available information. We need to anticipate all of the unanticipated events that could happen in how we design our infrastructure in the future.

There are going to be very difficult decisions in the future about how we go forward, and we need to start thinking and planning for those things now. That requires socialization and that requires a lot of careful thinking, and we need to do that.

Mr. DEUTCH. Thanks.

Senator, I hope that, in addition to taking these findings from this hearing back to Washington, that you will share these responses with the White House as well, as they are putting together this infrastructure plan. Everything that our witnesses suggested will not only help us address climate change but will help to preserve and strengthen our economy, which is the goal that I think we all share, Democrats and Republicans, and what is ought to be driving us today.

Senator NELSON. I am curious, Dr. Berry. What do you mean by a road—I got the impression that it is a road that absorbs water?

Dr. BERRY. Yes.

Senator NELSON. Tell me, what is it made of?

Dr. BERRY. I know in Chicago that any sidewalk renovation is done with permeable sidewalks. This is not my field, so I am kind of brainstorming. But can we have roads that do not spew water everywhere and create new problems on either side, particularly here in Florida? But if we had roads that could absorb water rather than disperse water, then I think we would be dealing with two problems at the same time. I know there is some technology.

Senator NELSON. You are seeing that in driveways, for example, in homes. The intrigue that you have caused me to wonder is, can you get that kind of surface that will stand up to the wear and tear of trucks, et cetera?

Dr. BERRY. That is where our challenges come from.

Senator NELSON. That is exactly right.

Madam Mayor?

Ms. MUOIO. Just to sort of add to that, I think we have to be rethinking our whole transportation system, and roads and cars and be working on mass transit and alternative ways to get people around other than building another road.

So a question I have as a local leader, we are charged locally with building codes and establishing building codes. What would you advise us to do in terms of our building codes to be more resilient to sea-level rise? What kinds of things should we be including in our building codes moving forward?

We can start with Dr. Kirtman.

Dr. KIRTMAN. So, first of all, I think all building codes should include all the possible scenarios of sea-level rise over the lifetime of the building. So that should be a requirement. We should be thinking about how does a building recover from tidal flooding and storm surge flooding.

So I think we have to think that there is going to be time when the ground floor, whether that be parking or what, is going to be flooded. And how rapidly are you going to recover?

So I think we have to think that our buildings are going to have a certain amount of flooding that is going to happen, but how quickly can you recover? I think that is how we have to think about our building codes, is think about rapid recovery from anticipated events.

Ms. MUOIO. Dr. Berry?

Dr. BERRY. Already, in Miami-Dade, a number of hospitals have taken electrical equipment and raised it. I think that is for houses too. I think we have to look at the potential fire hazards and potential electrical hazards and make sure those are out of a reasonable flood zone.

So that is not a usual building code, but that is one I think ought to be implemented.

Mr. HEDDE. I would identify three areas.

I think from the local community standpoint, it starts with a planned use plan and keeping structures out of harm's way. That is one of the easiest things.

As I have worked on building code issues, I have been appalled at the lack of strength in some communities' building codes. Luckily, in Florida, they are fairly strong.

So the second piece I would agree would be the elevation of the structure, make sure the structure is elevated.

I think the last piece that people forget is that we need to build foundation systems that are strong enough to withstand storm surge. A lot of these things are not strong enough. So in spite of being elevated, they do not withstand the storm surge.

Dr. JURADO. I think that the responses thus far have been quite substantial. I appreciate the need to elevate structures and have some uniformity. Consistently, we hear about the importance of having consistency across our codes, making sure that whatever we do in the build environment will be consistent with transportation and drainage systems, since everything is interconnected.

But independent of the resiliency of flood protection, focus I think on the challenge in Florida, which is the inability to dictate really on how to treat energy within the build environment. There are substantial limitations about requirements for renewable energy many local governments, for example, might want to see advance.

So some flexibility with regards to energy code at the State level would be useful.

Ms. MUOIO. Very good. Thank you so much for your comments. I appreciate it.

Senator NELSON. Madam Mayor?

Ms. BURDICK. Thank you.

We have talked a lot about what we can do to mitigate the impacts of climate change. And I thought I heard, Dr. Berry, that there is nothing that we can do to mitigate those in the near future?

We talk about carbon and carbon tax. We talk about trying to reduce the problems of climate change. We hoped that we were going to get cutbacks on coal, and it looks like some of that is changing.

So what are some of the preventive things that we should be doing locally, nationally, and globally, to reduce climate change? And how do we best educate our public?

Dr. BERRY. For Florida, I am not a pessimist, but I am a realist, in the sense that we should mitigate what we can. We should have clean energy, and we should be efficient. But the rest of the world is imposing its will on us at least for the next 20, 30, 40 years in that the emissions that we are already on our way to creating or are already there are going to result in sea-level rise. If we stop everything now, there still will be sea-level rise. We are not going to stop everything now. It is too complicated.

So sea level and emissions are going to go on in the best world, in my opinion, for 20 or 30 years. Therefore, given where we are, we have to mitigate the impact of the global sea-level rise.

At the same time, we ought to be moving to new forms of energy. We ought to be moving, as the mayor says, to thinking through not just electric cars but ways of transportation. We started. We have local communities that are very advanced in that.

Changing our ways of life individually is something that I do not do very well but some people do in terms of where we go, what we transport, how we manage driving a single car or driving with four people. There are lots of small things that we ought to do. But we also ought to be realistic and say the big picture for South Florida is not going to change until we better understand how we can manage it and take advantage of it.

What I say in my written testimony is that we have the opportunity, as Commissioner Abraham said in another interview, of creating innovation technology in dealing with flooding, in dealing with structure that I have talked about.

So we can turn a problem into an opportunity here in South Florida. We are the bad example, if you like. Bad, but we can turn it into something good. It is a lesson, like the Dutch. The Dutch are exporters of technology to deal with floods.

We could create our own patterns of technology, our own social responses, and, therefore, be exporters of good things to the rest of the coastal areas of the country that are going to have the same problems that we are having maybe 5, 10 years down the road.

Ms. BURDICK. It makes that collaboration piece with the South Florida regional climate compact all the more important so that we can all work together to find these innovative solutions and implement them.

Thank you.

Senator NELSON. And the big difference between the Netherlands and Florida is that we sit on top of a honeycomb of all kinds of materials that are decayed from seashells and critters over millions of years, and that honeycomb of limestone is filled with water. Thus, saltwater intrusion, because saltwater is heavier than freshwater. As the saltwater rises, it comes inland, and it invades our honeycomb.

You, Dr. Jurado, have had a number of your wellfields in Broward County invaded. You have closed two in the county. Hallandale Beach has relocated further to the west six. Deerfield Beach has abandoned eight.

So what in the world?

Dr. JURADO. The issue of saltwater intrusion is incredibly problematic. Working with the USGS, we have seen that sea-level rise has accelerated the land movement of the front by about a factor of two. We, as I indicated, expect to lose about 40 percent of coastal wellfield capacity in the future sea-level rise, and that is sea-level rise, because we are actually precluded, the regional policy for taking more from the Biscayne aquifer, and thus we have an offset. So this was really driven by sea-level rise.

And so that will require substantial relocation of wells to the west. We are also working with Palm Beach County, however, and a diverse number of stakeholders on a regional reservoir. And that is the other thing that we need in our region is more storage, because we simply lack the land for long-term water storage. Even though we receive rainfall of 20 inches at a time, we are trying to find a place to put it.

But storage in the system when we are trying to manage saltwater intrusion in extreme drought is going to be really critical, and it is another piece of infrastructure funding. We are pursuing funds to the tune of $160 million for Phase 1 of the C–51 reservoir, which could provide water supply and environmental benefits for Palm Beach, Broward, and Miami-Dade counties—full-scale, a $460 million reservoir providing up to 150 million gallons per day and water supply for the region.

So investments in alternative water supplies such as storm water reuse and reclaimed water coupled with conservation will be an absolute necessity.

Senator NELSON. And, Congressman, this fits hand-in-glove with our Everglades restoration efforts, moving that water south and increasingly getting it cleansed as it is coming south, ultimately into Everglades Park into Florida Bay as well as the Shark River Slough.

How do these municipalities monitor the aquifer to determine whether the freshwater has been contaminated?

Dr. JURADO. We have an extensive groundwater monitoring network with wells that are maintained by the South Florida Water Management District, by the individual water utilities, by Broward County, and by the USGS. So this saltwater network is monitored monthly. We have more than 50 years of data the state has now used to help develop models that are informing our decisions today.

So it is a vast network, one that we are actually upgrading. Just now we have been holding regional workshops. Since the saltwater front has moved beyond many of the wells that were previously used to measure the advancement of the front, they are already salty, so we are having to recalibrate that network for the current condition. But it is aggressively monitored and wellfields managed in accordance with these water shortages and predictive losses.

Senator NELSON. And moving those wells further west ultimately ends up with the customer paying more because of the added expense.

Dr. JURADO. That is correct. Moving the wellfields will cost something. We also have to recognize that we have water treatment facilities that are associated with individual wellfields. And if we have to move the wellfields outside of the regional water treatment facilities, then you are potentially looking at new water treatment facilities as well.

The other thing though is, if we have to move to reclaimed water or brackish water sources as an alternative water supply, those require more in the way of treatment to produce potable water and much higher energy costs. So the reservoir option is highly attractive because it is freshwater. It does not require a higher level of treatment, such as the removal of salt.

So storm water reuse coupled with conservation are really the most attractive strategies.

Senator NELSON. Dr. Kirtman and Dr. Berry, you know I mentioned at the outset the attempts to try to cut the budgets of anything that measures climate change. This is all the way from programs existing in NOAA and NASA to future satellites that would have all of these very intricate measurements.

Now, to what extent does your work rely on climate-observing platforms like those satellites?

Dr. KIRTMAN. A great deal of my work relies on observing networks.

So one of the projects we have at the University of Miami is to make forecasts based on decades. And how do you make those forecasts? You need to have some way of saying what the state of the climate system is today in order to project it into the future. That information, that state of the climate system today, comes from our observing networks, our satellite platforms, our ocean buoy systems, our drifter buoy systems, our radio sound data.

If we stop collecting data, that is truly putting your head in the sand. Just because you stop collecting data does not mean the climate change problem is not continuing to progress.

So first and foremost, you need to monitor the health of the climate system. That is first and foremost. We have to put a red line in the sand in stopping any attempt to remove the collection of data. This is critical.

Senator NELSON. Just to put a point to this and to underline it, I want, Dr. Kirtman, if you will, just shortly describe how a lot of this data that you are talking about for climate change also goes into our computers to help us with the direction and the velocity and the intensity of hurricanes.

Dr. KIRTMAN. Exactly. That is a wonderful question, Senator.

It is exactly the same kind of data that we use to make better forecasts for extreme rainfall, better forecasts for hurricanes. All of that data is the data of the climate system. We do not make a distinction when we observe the atmosphere of the ocean, the land surface. We do not make a distinction whether it is climate change data or whether it is just weather data. It is data.

If we throw away that data on the state of the ocean because we worry about it revealing some climate change, we are throwing away our opportunity to improve our forecasts for the intensity of hurricanes, our ability to predict storm surge.

I cannot say enough how much of a threat this is to lives and property and economic opportunity and national security. This is day-to-day weather, not only climate change. It is of paramount importance.

Senator NELSON. Dr. Berry, did you want to add to that?

Dr. BERRY. I am as passionate as he is with this. It is not just for science. Communities that are dealing with current problems need to have information. Miami Beach cannot put the pumps in there and just guess that sea-level rise or that king tides are a certain height or sea-level rise is going to be something. We need information. Companies need information. Engineering work needs that same information.

The company I work with at the household level is using LIDAR data based on NOAA. We are using tide data. We are using all kinds of data that come out of the Federal Government.

This government, I mean, I was born in England, and one of the things that I recognized many decades ago is that information is treated differently in different parts of the world. We have a tradition going back dozens, hundreds of years of saying the data comes from taxpayers and it goes back to taxpayers. It is open.

And I was so impressed with the USGS at one time putting data out on streamflow. And who used it? Trout fishermen, OK? And that is great.

And so I cannot emphasize how important it is not just for science but for everybody.

Senator NELSON. And that is so true in Florida not only for commercial fishermen but for recreational fishermen as well.

Now, Mr. Hedde, your industry, the reinsurance industry, so you are insuring against catastrophe, you really rely on this data, don't you, to make your calculations?

Mr. HEDDE. Yes. So I asked if I could comment because we definitely rely on this information.

That is a surprise to some who really do not understand our industry, where we rely on this information to value risk.

When you value risk, if you add uncertainty into the value, that impacts the valuation, usually on the higher side of the price. So that is one way.

A lot of our transactions, especially in the reinsurance industry, are based on triggers that are based on NOAA data that's supplied. So I made a point in my testimony today to say that we rely on NOAA information.

Senator NELSON. All right. Just so that you will know how we are getting all the more accurate, you have heard of the Hurricane Hunters. They fly into the hurricane. They are usually in the range of 15,000 to 30,000 feet.

They are taking data real-time from their instruments. They are dropping an instrumented package about that long. It has a little parachute on it, and it goes all the way down to the ocean. It sends back real-time data to the airplane, which is sent immediately by satellite to the National Hurricane Center.

At the same time flying over the top of the hurricane is the G–IV. We had it down one time during a hurricane, so I think we are able to get another one. We just got it in the NOAA bill. And it is doing the same thing, from 45,000 feet, dropping an instrument. Here is the latest gizmo, and the P–3 that goes into the storm.

Now they are putting out through a hole that big an instrumented package that comes out and, all of a sudden, it sprouts wings and a motor fires up. And now you have a UAV inside the hurricane that can fly actually in and stay in the hurricane eyewall giving us more measurements.

And they are thinking that we might get an accuracy as a result of that from 10 to 15 percentage points even more accurate in our data.

Now, Dr. Jurado, so the people that are trying to muzzle scientists—it is happening. I have seen it in Washington. I have seen it here in the state of Florida Government.

Could you do your job effectively if you could not use certain scientific terms?

Dr. JURADO. I cannot imagine planning for the future conditions of the community and making informed policy planning investment decisions without recognizing the reality of what the driving factors are. And in our region, climate change and sea-level rise are the primary influences for which we need to be organizing policy planning investments for decades to come. I cannot imagine those being informed or fruitful investments if they were not based upon credible use of scientific data and in large part on contemplation of the impacts of climate change on our environment.

Senator NELSON. And, Mr. Hedde, what would happen in your industry if you, as a consumer, a homeowner, could not get accurate climate data and climate change on which to make your decisions on what your purchases are?

Mr. HEDDE. We have seen a change in our industry over the last 5 years.

About 5 years ago, our clients would ask us to come and talk about climate change at the board level but tell us specifically not to use "climate change" in our words. Over the last couple years, we still come back and talk about climate change, but they are allowing us to use "climate change."

It comes back to my point before. We use the data to accurately assess risk. And insurance plays such an important part in our economy, if we are not properly able to assess the risk, it is a drag on the economy.

So it makes our answers better. It makes our product better. It makes our response to events better.

Senator NELSON. And would you believe that I have even had to file a Scientific Integrity Act to define common scientific values free from political interference?

Mr. HEDDE. It does not surprise me.

Senator NELSON. It has come to that, as well as to try to keep from the intimidation and censorship that we see.

Congressman, do you have another question?

Mr. DEUTCH. Thank you, Senator.

I just want to follow up on your important question about terminology and what is actually happening.

I think it is fair to say, and the witnesses can correct me if I am wrong, and, Mr. Hedde, I think you just alluded to this, that every major company in the United States and likely in the world devotes some core part of its time and resources to addressing the issues that we will be grappling with as a result of climate change, whether it is supply chain disruptions, issues driving sustainability, the severity of storms.

So the business community understands that we have to be tackling this, and they do it in a thoughtful and responsible way. And I am sure the same is true for Senator Nelson for meetings I have had in my office with corporate leaders from some of the largest corporations, including energy companies who understand that we have to start to anticipate the impact this is having, address it, and even recognize that carbon has a cost in doing business.

So the business community gets it.

Local government, again, Dr. Jurado, as a professional, not a politician, local governments run by Democratic and Republican administrations who are grappling with sea-level rise firsthand and the impact firsthand, and the issues of climate change firsthand, are having responsible discussions about how to address it, because as Dr. Kirtman points out, when we have multiple lines of evidence making it clear that human activity makes a dramatic impact on climate change, then we all have to be examining it.

Why is it—I am not sure that anybody can answer this. Perhaps it is a rhetorical question. But why is it that when the business community understands what is at stake for the future of our economy and our country, and our local governments understand what is at stake, why is it that there are some in important roles in our government who continue to, as, Dr. Kirtman, I think as you put it, continue to put their hand in the sand? And, by the way, in Florida, that sand is the coastline and the water is getting closer.

[Laughter.]

Mr. DEUTCH. So why is it that there is this ongoing denial that we see in some very narrow segments of our government?

I am not sure, Senator, if anyone wishes to answer that.

[Laughter.]

Mr. DEUTCH. But I suppose I would ask another way.

You would agree that, given your eloquent testimony and the work that every one of you does every day in this area, that it is difficult if not impossible to accept the fact that there are those in important positions of power who will continue to deny the necessity of taking on these issues that every one of you has devoted your life to?

Four nods across the table.

Senator NELSON. I am going to ask the Congressman's question in another way.

First of all, one of you testified about, particularly after Hurricane Andrew, the changing building standards.

That is something I will never forget, flying in a National Guard helicopter over ground zero the day after Hurricane Andrew. There were two buildings that were left standing. One was the bank building, and the other one was an old Florida cracker house that had been built in the 1920s to withstand wind. Everything else was leveled.

Indeed, as a result of that rather traumatic experience in 1992, we have really, to the credit of the local communities and their governments, brought up building standards.

So, certainly, that means a lot to you, Mr. Hedde. But it also means a lot to folks like this from local government that you do have predisaster mitigation grants and community rating systems to reduce the risk. And that means that you need those FEMA programs.

And remember what I said? They were reducing FEMA 11 percent in the budget for these kind of programs? What would that do to you in Broward County, Dr. Jurado?

Dr. JURADO. Broward County and many of the local communities in our state are very active participants in CRS. The CRS program has served as a critical incentive for really robust flood mitigation strategies that reduce our reliance on FEMA because we have built-in protections in the way we develop and design the community.

The savings are, I don't know, in the tens and hundreds of millions of dollars in terms of the policy reductions that are delivered to residents in the community. We are strong supporters of the CRS and the type of robust standards that have elevated all of our flood protection measures by helping deliver resources to communities.

And with FEMA being $30 billion, $20 billion in the negative, we can no longer afford to expose ourselves to flood risks, and we need programs that help all of our communities build resilience and do it cost-effectively. The cost of insurance has to be something we keep our eye on.

Senator NELSON. Mayors, I have one more question. Do you have any questions?

Ms. MUOIO. Thank you so much for being here.

Senator NELSON. Absolutely. Thank you, Madam Mayor.

All right, Mr. Hedde, you mentioned that there was over $40 billion in economic loss due to storm events just last year in 2016, and you said that about half of that was insured losses. So that means the remaining half was not insured, so that was going to have to be paid by somebody.

So given the modeling scenarios that you have seen, do you think those losses that are primarily being borne by the individuals, since they were not insured losses, will rise in the future, given what we are seeing?

Mr. HEDDE. That is a tough calculation. We see a pretty steady historical insured versus noninsured penetration within the U.S., so I think everything will rise most likely at that variable.

What disturbs me more is the lack of penetration on flood insurance, where we do not see a large penetration of flood insurance or earthquake insurance. I think that ultimately will impact the banking industry, because so much of their portfolio is not insured upfront.

So we are looking at solutions for increasing the penetration of flood insurance and participating heavily in NFIP program insurance this year. We developed our inland flood product that we are offering to people, not competing with the NFIP but supplementing people who do not live in coastal regions but are subject to inland flooding.

So I think there are ways to address that. Ultimately, I look at it as we need to change behavior, both in the long term and the short term.

And coming back to the question around funding, I think we need to make incentives available to help people change behavior, become more resilient, whether it is building a stronger house, whether it is energy efficiency. There are so many things that we need to have funds for to be able to take care of this problem.

Senator NELSON. I want to thank everybody. We have covered this subject in depth. It is exactly what we wanted to do in a field hearing.

Thank you, and the meeting is adjourned.

[Applause.]

[Whereupon, at 3:14 p.m., the hearing was adjourned.]

APPENDIX

RESPONSE TO WRITTEN QUESTION SUBMITTED BY HON. BILL NELSON TO
BEN KIRTMAN, PH.D.

Question. Dr. Kirtman, you told us that for about 800,000 years, the carbon dioxide levels in the atmosphere stayed between 180 and 280 parts per million. And in less than 150 years, we've seem an unprecedented rise in CO_2. A few years ago, I visited the NOAA observatory on top of Mauna Loa, and it was a bittersweet moment. The CO_2 reading was 399. The scientists told me it was probably the last reading that would be below 400 parts per million. Today, we are at 405 parts per million.

Dr. Kirtman, without action to curb emissions, what is the likely carbon concentration we will see in 10 years? And even with significant climate action now, will we still need to invest in adaptation to make coastal communities more resilient?

Answer. Given the current trajectory, I would expect CO_2 levels to go from 405 ppmv today to about 450 ppmv ten years from today.

Even if we were somehow about to reduce CO_2 levels to say 370 ppmv (the concentration from 2000) today we would still be committed to significant warming for the next 25 years. This is the climate change commitment that is already "baked-into" the system. Essentially, adaptation will always be a 25-year or so time horizon challenge as long as greenhouse gases in the atmosphere continue to rise.

RESPONSE TO WRITTEN QUESTION SUBMITTED BY HON. BILL NELSON TO
LEONARD "LEN" BERRY, PH.D.

Question. Dr. Berry, your testimony included a letter to the President signed by several Florida scientists that notes three important climate change considerations: the need for continued Federal earth science research, the importance of scientific integrity, and federally important coastal properties at risk.

The letter specifically mentioned the Kennedy Space Center and Cape Canaveral. Can you describe how sea level rise may threaten our space launch capabilities?

Answer. In highlighting the vulnerability of many military facilities to sea level rise, we, as scientists, are stating the obvious.

Air Force bases, Naval bases, and many Army bases are located in coastal areas, and in the search for flat, available land, many are on the coast at low levels above the high tide mark. In a world where sea levels had remained relatively stable for hundreds of years, this made sound sense, particularly as aircraft and rockets could take off across the ocean.

In the world of today, where sea levels have already risen at least nine inches, and are on an accelerating course, with anticipated rises of up to two feet in the next thirty years, and more than double that by the end of the century, those coastal locations can present tough problems.

Higher sea levels provide a higher base level for storm surge and also high ground water tables which can compromise infrastructure. Bases in Florida provide plenty of examples. Patrick (near Cocoa), McDill (near Tampa) and Elgin (in the Panhandle) Air Force Bases all have areas close to sea level. McDill has a mean height of about five feet, easily vulnerable to storm surge now and in the future. Patrick AFB and Elgin AFB are higher but have significant areas of vulnerability.

But the most threatened facilities are at Cape Canaveral, where a recent report has said "rising sea levels are the single biggest threat to the facility with over two thirds of constructed infrastructure and more of the land area are below sixteen feet and vulnerable to storm surge as sea levels rise." (Source: *https://www.giss.nasa.gov/research/features/201508_risingseas/*)

It is important to note two things. First, the bad news is that projections of the rate and amount of sea level rise are growing steadily, as we have more information and understanding of the processes involved. The second is better news, that with ongoing basic research and analysis, we can be continually better informed and able to adapt and protect these vital facilities.

As many have said, adaption needs to start now based on the best available science and the best ongoing collection of vital data. The very facilities under risk are those needed to preserve our information flow and our ability to guide our future.

RESPONSE TO WRITTEN QUESTIONS SUBMITTED BY HON. BILL NELSON TO CARL G. HEDDE

Question 1. Mr. Hedde, you noted that most of our homes and businesses aren't built to withstand the severe weather we currently experience from time to time, much less more extreme events in a changing climate. Is this true for hospitals and schools too?

Answer. It is not possible to make a blanket statement regarding the construction quality of our Nation's schools and hospitals. Engineers are required to incorporate safety features based on a building's primary or intended function. For example, ASCE 24–14 is the industry consensus standard for flood resistant design and construction, and uses four building categories to guide engineers on flood design specifications: the first is for accessory buildings (*e.g.,* garages), the second is for residential homes and general commercial buildings, the third is for large gathering places, and the forth is for essential service facilities, such as hospitals and fire stations. In ASCE 24–14 elementary schools are considered a class three building, unless they also serve as a designated emergency shelter, in which case they are considered a class four. The higher the class, the more the building must be elevated and flood-proofed. Similarly, seismic loads and design requirements are more robust for schools and hospitals. Wind loads are typically increased for schools and hospitals either through an importance factor in older versions of codes and standards, or through maps with higher design wind speeds in more recent building codes and standards. While ASCE 24–14 and other wind and seismic standards have been around for decades, not all buildings were historically built to code, while others may have been constructed after a code was put into place, but lax enforcement allowed the building to be under-constructed. We have seen numerous examples of Hurricanes and Tornadoes causing severe damage to schools and hospitals. As a nation, it is important that infrastructure such as schools and hospitals be constructed based on forward looking conditions that might be experienced in the future.

Question 2. This August will mark 25 years since Hurricane Andrew devastated Homestead, Florida. If Hurricane Andrew were to happen in 2017, given our current building codes and coastal development patterns, what would the loss be like?

Answer. Hurricane Andrew was a strong wake up call for the need for stronger building codes and practices. More than half of the U.S. population lives in one of the Nation's 673 coastal counties, and values of properties in these coastal regions have increased since Hurricane Andrew. If a storm similar to Hurricane Andrew would happen in 2017, the resulting damage would be heavily dependent on where the storm occurs. While building codes have improved during the years since Hurricane Andrew, building code application and enforcement varies by state, and further varies by local community within states. Since Hurricane Andrew, building codes in Florida have been greatly improved, and we would expect better building performance for buildings build in Florida's coastal regions since the post Hurricane Andrew building codes were implemented. Unfortunately, the same strengthening has not occurred in all coastal regions along our Nation's coast. Organizations such as IBHS (Insurance Institute for Building and Home Safety) have conducted extensive research on building performance, with the ultimate goal to improve building practices and products, and better inform the development of stronger building codes.

Question 3. This August will mark 25 years since Hurricane Andrew devastated Homestead, Florida. Are there resources available to homeowners who want to strengthen their homes, and have we supported those efforts sufficiently at the Federal level?

Answer. A number of insurers and reinsurers writing insurance in the United States have formed and financially support a non-profit organization dedicated to researching and communicating effective measures that can be taken to reduce loss to human life, homes, and businesses from natural disasters. IBHS (Insurance Institute for Business and Home Safety) conducts critical research that will improve building practices, building materials and building codes. Insurers continue to new find ways to communicate with policy holders on ways to protect and strengthen their homes and businesses. Munich Re has developed The FORTIFIED Home™ On the Go Tablet App. The app has been designed for an iPad, is an interactive tool for homeowners, contractors and architects to use when building or retrofitting a single-family home to make it a more fortified structure. It includes information about the Insurance Institute for Business & Home Safety (IBHS) FORTIFIED Home™ program which is designed to strengthen homes. The "app" is a "free" app available on the iTunes app store. We would ask that all Congressional offices make the "app" available to their constituents.

The Federal Government is exploring several new programs to help prioritize risk mitigation spending over disaster response. The FEMA Disaster Deductible is a good example of risk sharing and incentivizing mitigation. FEMA is financially supporting a new study to update the often stated statistic that every dollar spent on mitigation saves $4 is disaster recovery. The study will be completed by the National Institute for Building Science (NIBS). We feel that the study should be fully funded by Congressional appropriators.

RESPONSE TO WRITTEN QUESTION SUBMITTED BY HON. BILL NELSON TO DR. JENNIFER L. JURADO

Question. Dr. Jurado, you mentioned that a recent estimate done in partnership with the Corps of Engineers suggested that for Broward County to execute a plan for resilient seawalls would cost anywhere from $1 billion to $4 billion—and that failing to act would be far more costly. How can counties like yours even begin to finance such massive capital improvements?

Answer. Just to clarify, Broward County is in the midst of a study with the Corps of Engineers to develop recommended sea wall top elevations needed to deliver flood protection with rising seas. The study will formally evaluate the flood protection level of service derived with a 2.5 foot increase in sea wall height. Concurrently, Broward County is developing cost estimates for collective raising/armoring of the 350 miles of canal banks and waterways along our shoreline. The final figure will depend on the initial condition of the seawall and the construction approach, but we estimate the cost to be between $1 billion and $4 billion. This infrastructure upgrade will need to be coupled with additional stormwater improvements and resilient investments as part of development practices, efforts which are already underway.

Regarding the question of financing, local communities simply can't do this alone. The scale of investment needed far exceeds the resources and bonding capacity of individual communities, while the timeline for execution is urgent. A Federal partnership is essential, as has been the case in advancing many of the most sizeable infrastructure projects and investments in our nation—investments which have served to connect and protect our communities as they exist today, including ports, bridges, roadways, waterways, regional flood control systems, and reservoirs. These investments provided a foundation for communities, including our coastal communities, to develop and thrive. However, many coastal communities are now exposed to new pressures and excessive, unforeseen infrastructure needs, which threaten to overwhelm our capabilities. Local sources alone won't be sufficient, given the costs and timeframes involved. We are not punting responsibility; however, we recognize that diverse funding sources and alternative funding streams will need to be pursued. Therefore, we desperately need the support of the Federal Government, and of our Federal agency partners, including the Corps of Engineers, to undertake the large-scale flood protection investments our region needs. These infrastructure upgrades are vital to protect against catastrophic flood damage, service disruptions, and threats to public safety. They will deliver a sizeable return on investment, ensuring long-term economic well-being of national consequence, while substantially reducing the potential for costly Federal post-disaster assistance often required in the aftermath of severe floods and storms. Absent organized investments to address our risks, these events will only increase in severity and consequence.